W9-ADA-465

WITHDRAWN

544.926
L48i   76930

DATE DUE			
Feb 23 '72			

WITHDRAWN

IDENTIFICATION TECHNIQUES
IN
GAS CHROMATOGRAPHY

IDENTIFICATION TECHNIQUES IN GAS CHROMATOGRAPHY

D. A. LEATHARD

Department of Chemistry, University College of Swansea

B. C. SHURLOCK

Materials Division, Berkeley Nuclear Laboratories

WILEY—INTERSCIENCE

a division of John Wiley & Sons Ltd

LONDON   NEW YORK   SYDNEY   TORONTO

CARL A. RUDISILL LIBRARY
LENOIR RHYNE COLLEGE

544.926
L48i
76930
Dec.1971

Copyright © 1970 John Wiley & Sons Ltd.,
All Rights Reserved. No part of this publication may be reproduced, stored in a retrieval
system, or transmitted, in any form or by any
means, electronic, mechanical photocopying,
recording or otherwise, without the prior
written permission of the Copyright owner.

Library of Congress Catalog Card No. 75-122349

ISBN 0 471 52020 9

PRINTED BY UNWIN BROTHERS LIMITED
THE GRESHAM PRESS, OLD WOKING, SURREY, ENGLAND
A MEMBER OF THE STAPLES PRINTING GROUP

PREFACE

Methods of peak identification have advanced much more slowly than most other aspects of gas chromatography. Even today the majority of workers rely almost entirely on techniques involving retention coincidence, although they are almost certainly aware of the more highly publicised identification aids such as gas chromatography combined with mass spectrometry, and a few special detectors. As well as covering these areas, the major aims of this book are to make more widely known the many available but neglected techniques which lie between these extremes, and to try to bring some coherence to a subject which straddles many subjects and expertises.

The major periodical sources have been covered up to November 1969. Full titles of references are given as an aid to the selection of suitable further reading about particular identification techniques.

It may be worth asking whether qualitative gas chromatography is a sensible subdivision of the whole subject. We believe that it is, since although it overlaps in some ways with quantitative procedures and also borrows substantially from more conventional analytical techniques, it does involve many novel features, and represents an experimental corpus of great practical significance. Thus, if an apology must be given for another book on gas chromatography, it is that identification techniques have never before been fully treated in book form, that problems of identification by gas chromatography are widespread, and that there are many effective methods which are little known or exploited.

Wherever possible, we have not extensively developed topics which are treated fully in standard texts on gas chromatography. However, we have felt it necessary to devote a large chapter to the fundamentals of retention since these are often not fully appreciated in analytical circles and yet they form an important base for the understanding and effective use of retention coincidence techniques. Detectors are also considered in some detail since, although their selective powers are limited, their use demands little modification of normal procedures, and most laboratories are well equipped in this respect.

Inevitably, in a subject as diverse as this the selection of material and judgement of the relative importance of various techniques is likely to be controversial. In general, except for the last two chapters, we have assumed the viewpoint of the 'average' gas chromatographer who has limited resources and wants, from time to time, to solve difficult identification problems with the resources at hand. This does not mean, we hope, that those who study the subject *per se* and are involved in a great deal of peak identification will not find much of interest. We have not restricted ourselves to established methods but have pointed out new possibilities where we could see them.

The task of writing this book would have been considerably more difficult without the help of the numerous authors and colleagues with whom we have discussed many aspects of this broad field. We first saw clearly the need for the book while writing a review article on the subject at the suggestion of Professor Howard Purnell whose help, advice, and encouragement over a number of years are gratefully acknowledged. Special thanks are due also to Professor John Beynon, Dr. Harry Hallam, and Dr. David Locke for reading, respectively, Chapters Eleven, Ten, and Two. Any errors or misjudgements which remain are, of course, our responsibility.

We are grateful for the permission of authors and copyright owners to reproduce various diagrams and tables.

D. A. LEATHARD

B. C. SHURLOCK

CONTENTS

CHAPTER ONE

INTRODUCTION

Despite the extensive development of many practical and theoretical aspects of gas chromatography (GC), analysts are frequently still severely troubled by the presence of an unknown peak in a chromatogram. Indeed, most standard texts on GC give short shrift to methods of peak identification. Nevertheless, in recent years a substantial number of pertinent experimental methods and results, from diverse disciplines, have been published. There are several relevant review articles,[1-9] but none covers the whole field in any depth.

One of the striking things about peak identification is the wide variety of techniques available. Each approach has been devised to cope with a particular problem and may have a number of unique features, but many are capable of more general application. Most methods fall into one or other of the divisions which form the chapters of this book, although the boundaries between these are not always distinct (or meaningful) and some overlap is inevitable.

Identification problems in analysis range in severity from the mere confirmation of an almost certain identity to the characterisation of a sample about which virtually nothing is known. Most techniques have limits to the severity of problem to which they can be applied, while the particular approach adopted will almost certainly reflect the experience of the analyst and the particular facilities available to him.

GC is essentially a way of separating a mixture into its constituents, but at the same time three bits of information are obtained for each component: the retention volume, the peak width, and the peak height. Simple methods of identification are based on the retention volume and, much less frequently, on the peak width of a component. Although the disadvantages of methods of identification based on retention are now widely appreciated, they are still among the most commonly used. Indeed, they provide a useful means of confirming an identity which is strongly suggested by other considerations; but they are not suitable for analysing a completely uncharacterised sample. The attraction of identification by retention is that, in its simplest form, it demands no apparatus in addition to the chromatograph, which can therefore be used at the same time for both qualitative and quantitative analyses. However, rigorous identification from retention measurements is difficult, time-consuming, and ideally requires apparatus specially designed for simultaneous use of several columns.

1

A wide variety of methods of identification rely upon chemical treatments, ranging from class separation before injection to a variety of on-stream techniques. Among these, selective abstraction methods are frequently used, and provide a simple means of surveying the functional groups represented in a sample. Chemical modification techniques involve on-stream reactions of eluates, such as ozonisation, thereby multiplying the parameters by which a particular eluate can be identified. These methods, which rely upon versatile and imaginative application of established chemical procedures, almost invariably require a fair knowledge of the sample. In contrast, the more drastic methods of chemical modification such as pyrolysis gas chromatography are universally applicable. These techniques are analogous to spectrometric methods of identification in that they involve a 'fingerprint' approach.

Elemental analysis by GC may be achieved by breaking up the eluate into small characteristic molecules such as CO_2 and H_2O and analysing these fragments in a secondary chromatograph. By this means, in suitable cases, empirical formulae may be derived, and these, in principle at least, can be converted to molecular formulae by measuring the molecular weight of the eluate—possibly by one of the GC methods available. Qualitative elemental analysis can be carried out by on-line spectroscopic examination of the fragments of eluate produced in a microwave discharge or similar high-energy environment.

The specific response of a detector, or the relative response of two detectors, can sometimes be used to identify an eluate. Most detectors are well known to be unselective, but there are a few, such as the electron-capture detector, which are very useful in peak identification. Furthermore, comparison of the responses of two or more 'unselective' detectors can sometimes lead to a positive identification.

Standard analytical techniques can be adapted for the continuous monitoring of gas chromatographic effluents. For example, the Beilstein test, mass spectrometry, flame photometry, and thin-layer chromatography have all been used on-stream in conjunction with GC. Provided that reasonable sample sizes are available, almost any method of analysing the separated components can be used if peaks are trapped individually and then transferred to conventional equipment. While the conventional analysis of such trapped samples is beyond the scope of this book, the trapping methods themselves, which are of considerable importance, are described.

Although instrumental techniques, notably infrared and mass spectrometry, are potentially the most generally useful routine aids, they are also the most expensive, and are not always the best methods of identification.

Special problems arise in conjunction with the interface in directly coupled systems. Moreover, such systems usually involve the digestion of a great deal of data, and computing facilities quickly become a necessity.

The particular method of peak identification best suited to a given situation will depend very much upon the type of problem involved, upon the nature and size of the sample, and upon the apparatus and techniques at hand. Sometimes the use of more than one technique may solve a problem while none individually would do so.

Some rationalisation of the analytical problem may be gained by arbitrarily dividing it into three stages: guidance, substantiation, and confirmation. For example, selective abstractors might give guidance as to the type of compound present, while substantiation and further detail might follow from retention measurements, and confirmation of these details from consideration of other available factors such as detector responses and retention volumes of derivatives.

Some examples will indicate the scope and variety of methods available.

Chlorinated insecticides are frequently identified gas-chromatographically by retention coincidence, or by comparison of the responses from electron-capture and microcoulometric detectors. Comparison of the R_F values of the unknown and the standard from thin-layer chromatography also forms the basis of a method of identification: individual GC peaks may be deposited on the thin-layer plate and developed in the normal way. More recently[10] an improved technique of identification has been proposed. In this method the sample, dissolved in hexane, is irradiated with ultraviolet light and the resulting patterns of product peaks obtained by electron-capture GC are compared with those of authentic samples. Also, the p-values (equivalent to partition coefficients) between hexane and acetonitrile of the sample and degradation peaks are determined. On the basis of the observed extent of degradation, the pattern of degradation peaks, and the p-values of parent and degradation products, it is possible positively to identify a number of chlorinated insecticides.

Determination and identification of the components of flavour extracts from wines etc. is extremely difficult.[11] The human palate and the GC detector have very different specific responses, and the first problem is to pinpoint the GC peaks that are important in flavour determination. Identification of these peaks presents a major problem which can be solved only by using several techniques. Mass spectrometry is very useful but there are considerable limitations (see Chapter Eleven). A general guide to the types of compound present can be given by carrying out tests on the sample as a whole. For example, treatment of the flavour extract of Bartlett pear brandy with 2,4-dinitrophenylhydrazine indicates that it contains only a few compounds with a carbonyl group. Individual peaks have been identified by carrying out reactions on fractions separated by micro-preparative GC. Standard procedures such as hydrolysis, hydrogenation, oxidative and reductive degradation, etc., followed by GC analysis, enable a great deal to be found out about an unknown eluate.

Notation and terminology

In the following chapters the notation and some of the terminology of the American Society for Testing and Materials[12] have been adopted. The separated materials of a mixed sample are called 'solutes' while they are in the column, and 'eluates' when they emerge mixed with carrier gas from the column outlet in the 'effluent'. S.I. units have been used wherever they are unlikely to cause inconvenience and mental arithmetic. Thus, Hz is used instead of cycles per second, but mm Hg is retained in preference to Nm^{-2}.

REFERENCES

1. H. Boer, panel discussion on qualitative aspects of GC in *Gas Chromatography 1962* (ed. M. van Swaay), Butterworths, London (1962), pp. 316–19.
2. R. C. Crippen and C. E. Smith, 'Procedures for the systematic identification of peaks in gas-liquid chromatographic analysis', *J. Gas Chromatog.*, **3**, 37 (1965).
3. T. Gäumann, 'Gas chromatography', *Ann. Rev. Phys. Chem.*, **16**, 125 (1965).
4. G. Schomburg, 'Qualitative identification with the aid of gas chromatography', *Z. Analyt. Chem.*, **200**, 360 (1964).
5. D. A. Leathard and B. C. Shurlock, 'Gas chromatographic identification', in *Progress in Gas Chromatography*, Vol. 6 (ed. J. H. Purnell), Wiley, New York (1968), p. 1.
6. S. G. Perry, 'Peak identification in gas chromatography', *Chromatographic Rev.*, **9**, 1 (1967).
7. V. G. Berezkin and O. L. Gorshunov, 'Reaction GC in analysis', *Russ. Chem. Rev.*, **34**, 470 (1965).
8. R. L. Levy, 'Pyrolysis gas chromatography. A review of the technique', *Chromatographic Rev.*, **8**, 48 (1966).
9. W. H. McFadden, 'Mass-spectrometric analysis of gas-chromatographic eluents', in *Advances in Chromatography*, Vol. 4 (ed. J. C. Giddings and R. A. Keller), Marcel Dekker, New York (1967), p. 265.
10. K. A. Banks and D. D. Bills, 'Gas chromatographic identification of chlorinated insecticides based on their UV degradation', *J. Chromatog.*, **33**, 450 (1968).
11. E. Bayer, 'Quality and flavor by gas chromatography', *J. Gas Chromatog.*, **4**, 67 (1966).
12. ASTM Designation E 355–68, 'Recommended practice for gas chromatography terms and relationships', Philadelphia (1968).

CHAPTER TWO

PRINCIPLES OF RETENTION AND COLUMN SELECTIVITY

2.1 BASIC PRINCIPLES

Gas chromatographic theory can be split into two broad divisions. One aims to explain the shape and, in particular, the width of an eluted peak, while the other is concerned with the position of the peak maximum, i.e. the retention volume of the solute. It is not appropriate in this book to consider the origin of peak width in any detail, since only in special circumstances can the shape of a peak assist in its identification. On the other hand, the factors which influence retention must be considered, since it is common practice to identify solutes merely on the basis of their retention behaviour. In this chapter special attention is given to those factors which distinguish retention volumes on different stationary phases. As the details of GC become better known, it is becoming increasingly apparent that the processes involved are very complex. In what follows much detail and rigour has been omitted. Perhaps the most serious point which should be remembered is that in practice the retention of a component varies with sample size. This effect is usually minimised by working with the smallest possible sample sizes on columns with an inactive support material.

2.1a The fundamental retention equation

A gas–liquid column can be regarded simply as a volume of free space, V_G, with access to a volume of liquid, V_L. At equilibrium, solute molecules are distributed so that the ratio of concentrations in the liquid and gas phases is equal to the partition coefficient, K. Thus,

$$K = \frac{N_L V_G}{V_L N_G} \tag{2.1}$$

where N_G and N_L are the number of moles of solute in the gas and liquid phases. The probability of any solute molecule being in the gas phase is $N_G/(N_G + N_L)$, and in the liquid phase $N_L/(N_G + N_L)$. Hence,

$$\frac{N_G}{N_L} = \frac{t_G}{t_L} \tag{2.2}$$

where t_G and t_L are the total times spent by the average solute molecule in the gas and liquid phases, respectively. The total time spent by the solute in

5

the column is the retention time, t_R, given by

$$t_R = t_G + t_L$$

Together with equations (2.1) and (2.2) this equation leads to

$$t_R = \frac{t_G(V_G + KV_L)}{V_G}$$

Now, the volume of gas which flows from the column outlet during the retention time is defined as the retention volume, V_R. Hence,

$$V_R = F_o t_R = \frac{F_o t_G(V_G + KV_L)}{V_G} \tag{2.3}$$

where F_o is the flow rate at the column outlet measured at, or corrected to, the column temperature. For an incompressible fluid, $F_o t_G$ and the gas space of the column, V_G, can be equated. However, for an insoluble gas the actual volume measured at the outlet is larger than the free space owing to expansion of the gas throughout the column. It can be shown (see, for example, ref. 1) that, in these circumstances,

$$jF_o t_G = V_G \tag{2.4}$$

where j is a numerical factor smaller than unity and dependent only on the inlet (p_i) and outlet (p_o) pressures of the column; j is given by

$$j = \frac{3}{2}\left[\frac{(p_i/p_o)^2 - 1}{(p_i/p_o)^3 - 1}\right]$$

Reference 2 gives values of j to five figures for values of p_i/p_o between 1·001 and 3·000. Substituting equation (2.4) into (2.3) we obtain

$$jV_R = V_G + KV_L = V_R^0 \tag{2.5}$$

The left-hand term of this equation is called the 'corrected retention volume' and is usually written V_R^0, as shown.

All retention data are based upon retention times, or retention volumes measured at the column exit. Some of the measurements involved, and the relations between them, are given in Table 2.1. The notation used is based upon recent recommendations of A.S.T.M.[3]

2.1b Specific retention and free energy of solution

The value of a partition coefficient reflects the equilibrium distribution of a solute between two phases. We may write

$$\Delta G^0 = -RT\ln K = -RT\ln(C_L/C_G)$$

where ΔG^0 is the free energy of partition of the solute referred to a standard

TABLE 2.1. Retention parameters and measurements, and the relations between them

Name of parameter	Symbol and relations	Description
Retention time	t_R	Time elapsing between injection of solute and elution of its peak maximum
Retention volume	$V_R = F_0 t_R$	Volume of gas eluted during retention of the solute, measured at the temperature of the column and at the pressure of the column effluent[a]
Gas hold-up or air-peak retention time	t_M	Time elapsing between injection of insoluble gas (e.g. air) and elution of its peak maximum
Gas hold-up volume	$V_M = F_0 t_M$	Volume of gas eluted during retention of an insoluble gas, measured at the temperature of the column and at the pressure of the column effluent
Adjusted retention volume	$V_R' = V_R - V_M$	Volume of gas eluted between the air and solute peaks, measured at the temperature of the column and at the pressure of the column effluent
'Dead space' of column	$V_G = j V_M$	Total gas space of the column[b]
Corrected retention volume	$V_R^0 = j V_R$	Retention volume corrected for the compressibility of the carrier gas

TABLE 2.1. (*cont.*)

Name of parameter	Symbol and relations	Description
Net retention volume	$V_N = V_R{}^0 - V_G$ $= j(V_R - V_M) = jV_R'$ $= jF_0(t_R - t_M)$ $= KV_L$	That part of the corrected retention volume which can be attributed to solution processes[c]
Specific retention volume	$V_g = V_N(273\ °K)/w_L\,T$ $= \dfrac{jF(t_R - t_M)\,(273\ °K)}{w_L\,T}$	Net retention volume per gram of liquid phase corrected from the temperature of measurement of the flow rate to 0 °C[d]

[a] F_0 is the flow rate measured at the column outlet at the temperature of the column, T. If flow rates are measured at some other temperature, T', then $F_0 = F(T/T')$, where F is the flow rate at T', with temperatures in °K.

[b] j is the gas compressibility correction factor.

[c] V_L is the volume of the liquid phase on the column.

[d] w_L is the total weight of liquid phase on the column. Note that, although V_g is a function of the column temperature, T, it can be calculated in terms of the flow rate, F, measured at any other temperature T' (see footnote a).

state for the solute of unit concentration in the gas and liquid phases, and C_G and C_L are the concentrations of the solute in these phases. It is more usual in solution studies to work with standard states of 1 atmosphere pressure in the gas phase, and unit activity in the liquid phase, with activity defined so that at infinite dilution the activity of the solute is equal to its mole fraction. Hence, we consider a different partition coefficient, K_T, given by

$$K_T = a/p$$

where a is the activity of the solute in the liquid phase and p its partial pressure in atmospheres in the gas phase. In GC, conditions are almost always close to infinite dilution, and so by definition a may be replaced by x, the solute mole fraction. Thus,

$$K_T = x/p = KM/\rho_L RT \qquad (2.6)$$

where ρ_L and M are, respectively, the density and molecular weight of the liquid phase. Finally we may write

$$\Delta G^0{}_{\text{soln}} = -RT\ln\left(KM/\rho_L RT\right)$$
$$= -RT\ln\left(V_N M/w_L RT\right) \qquad (2.7)$$
$$= -RT\ln\left(V_g M/273R\right)$$

where the free energy of solution, ΔG^0_{soln}, refers to solute standard states of 1 atmosphere in the gas phase and unit activity in the liquid phase. This series of equalities demonstrates the equivalence of K, ΔG^0_{soln}, and the specific retention volume V_g (Table 2.1) as characteristic parameters of retention; any one of them completely characterises the solution behaviour of a given solute on a given stationary phase at a given temperature, independent (in principle) of percentage loading, inlet pressure, column length, or flow rate.

Determination of the partition coefficient K for a particular system requires an accurate knowledge of the volume of liquid phase present. This can only be calculated from the weight of liquid phase if its density is known at the column temperature. The specific retention volume V_g therefore has advantages as a characteristic physical parameter, since it can be determined from retention data, given only the weight of liquid phase, which is independent of temperature. For largely practical reasons most workers have virtually ignored V_g in favour of *relative* retention data. This is probably fortunate since estimation of the weight of liquid phase on the column invariably involves significant errors, especially at low loadings or when column bleed is significant. Various sources of non-ideality associated with the use of finite sample sizes and with the chromatographic process itself, which require elaborate correction procedures, also make it unrealistic to accord very great accuracy to values of K or V_g determined from routine GC work.

The use of relative retentions (see later in Section 2.2) removes some inaccuracy and also eliminates the need for making the j pressure-drop correction. However, the equations given above are implicit in the use of all relative retention studies. In particular, they demonstrate that a relative retention value measured on one column will only apply to another with the same liquid phase if ratios of net, adjusted, or specific retention volumes, V_N, V_R', or V_N, are used (see Table 2.1). These parameters implicitly involve correction for gas hold-up.

One of the advantages of dealing with ΔG^0_{soln} [equation (2.7)] is that it is usually related in a simple way to molecular structure, showing, for example, approximately constant increments for successive members of a homologous series. An approximate value of ΔG^0_{soln} can even be obtained by simply adding together the contributions of the various groups in a molecule. Indeed, this fact lies at the root of the use of retention indices and other correlation techniques for identification purposes. The free energy of solution, being directly related to retention parameters, is therefore a useful concept for considering retention data as a basis for identification.

On a particular column,

$$\Delta G^0{}_{\text{soln}} = \text{const} - RT \ln V_N$$

Hence,

$$\Delta G^0{}_{\text{soln}(2)} - \Delta G^0{}_{\text{soln}(1)} = RT \ln \frac{V_{N(1)}}{V_{N(2)}}$$

so that the logarithm of a relative retention value, when multiplied by RT, gives the difference between the free energies of solution of the two solutes. This observation is taken further in Chapter Three where the various methods of reporting retention data are discussed in greater detail.

2.1c   Vapour pressure and the activity coefficient

For an ideal solution the term p/x in equation (2.6) is equal to the saturated vapour pressure P of the pure solute (Raoult's law), while for non-ideal solutions at infinite dilution, $p/x = \gamma^0 P$, where γ^0 is an activity coefficient. Thus, from equations (2.6) and (2.7) we obtain

$$V_g = 273 R / \gamma^0 P M \tag{2.8}$$

and

$$\Delta G^0{}_{\text{soln}} = RT \ln P + RT \ln \gamma^0$$

Now, the standard free energy change of vaporisation of the solute is given by

$$\Delta G^0{}_{\text{vapn}} = -RT \ln P$$

$$= \Delta H^0{}_{\text{vapn}} - T \Delta S^0{}_{\text{vapn}}$$

where $\Delta H^0{}_{\text{vapn}}$ is the enthalpy change of vaporisation, and

$$\Delta S^0{}_{\text{vapn}} \approx \Delta H^0{}_{\text{vapn}} / T_b$$

where T_b is the boiling point. $\Delta S^0{}_{\text{vapn}}$ is close to the Trouton constant, k, and therefore

$$RT \ln P \approx kT(1 - T_b/T)$$

For many substances k is close to 23 cal mole^{-1} °K^{-1}. Substitution of this value into the above equation leads to the approximate relation

$$\log_{10} P(\text{mm Hg}) = 7 \cdot 9 - 5 \cdot 0 T_b / T$$

The form of this equation is the same as that of the Antoine equation:

$$\log_{10} P = C - A/T$$

However, C is not equal to $7 \cdot 9$ for all substances, nor is A equal to $5 \cdot 0 T_b$, owing to deviations of k from the value 23. In general,

$$k = 4 \cdot 6(C - 2 \cdot 9) = 4 \cdot 6 A / T_b$$

and hence

$$\log_{10} P(\text{mm Hg}) = 2 \cdot 9 + 0 \cdot 22k - 0 \cdot 22k T_{\text{b}}/T$$

This equation, together with (2.8), leads to

$$\log_{10} V_{\text{g}} \approx \log_{10} \frac{273R}{M} + 0 \cdot 22k \frac{T_{\text{b}}}{T} - 0 \cdot 22k - 2 \cdot 9 - \log_{10} \gamma^0 \qquad (2.9)$$

It is clear therefore that, unless γ^0 is very different from unity and shows variations from solute to solute, identification on the basis of retention data is essentially equivalent to identification by boiling point (see Figure 2.2). Thus, γ^0 is the thermodynamic measure of column selectivity.

2.2 THEORY OF COLUMN SELECTIVITY

Identification from retention more frequently depends upon relative, rather than absolute, retention volumes. Hence a factor of great importance is the relative retention of two solutes on the same column, r_{12}.

We shall call a solvent 'selective' towards a particular solute type if retention volumes for such solutes are significantly different from those expected on the basis of vapour pressure alone, i.e. $\gamma^0 \neq 1$ [see equation 2.8)]. In the present context, the term 'selective' should not be confused with 'specific' since the latter implies a particular chemical interaction between solute and solvent.

As shown above, a measure of solvent selectivity is provided by the deviation of γ^0 from unity. The excess free energy of solution, $\Delta G^0_{\text{excess}}$, over and above that involved in liquefaction of the vapour, is $RT \ln \gamma^0$ [equation (2.8) et seq.], and therefore, with subscripts 't' and 'a' referring respectively to thermal and athermal terms,

$$\Delta G^0_{\text{excess}}/RT = \ln \gamma^0 = \ln (\gamma^0_{\text{t}} \gamma^0_{\text{a}})$$
$$= \ln \gamma^0_{\text{t}} + \ln \gamma^0_{\text{a}} \qquad (2.10)$$

where

$$\ln \gamma^0_{\text{t}} = \Delta H^0_{\text{excess}}/RT \quad \text{and} \quad \ln \gamma^0_{\text{a}} = -\Delta S^0_{\text{excess}}/R$$

The enthalpy or heat term $\Delta H^0_{\text{excess}}$ can be considered to arise because of differences in attractive or repulsive forces between molecules in solution and those in the pure solvent or solute, while the entropy term, $\Delta S^0_{\text{excess}}$, can be attributed to differences in size of molecules of solvent and solute. Such a separation of enthalpy and entropy terms is convenient, but is only approximate.

It follows from equations (2.7), (2.8), and (2.10) that the relative retention of two solutes on the same column is given by

$$r_{12} = \frac{V_{N(1)}}{V_{N(2)}} = \frac{\gamma^0_{t(2)} \, \gamma^0_{a(2)} \, P_2}{\gamma^0_{t(1)} \, \gamma^0_{a(1)} \, P_1} \tag{2.11}$$

where the subscripts are self-evident. Since the ratio of saturated vapour pressures is fixed at any temperature, variations of r_{12} from column to column can only depend upon variations of γ^0_a and γ^0_t. It will now be shown that of these, γ^0_t, the thermal component of the activity coefficient, is usually the more critical factor.

2.2a   Excess entropy of mixing

The entropy term γ^0_a depends on the relative molar volumes of solute and solvent, v_1 and v_s respectively. According to the Flory–Huggins theory,[4]

$$\ln \gamma^0_a = 1 - \frac{v_1}{v_s} + \ln \frac{v_1}{v_s} \tag{2.12}$$

The Miller–Guggenheim theory[5] leads to

$$\ln \gamma^0_a = 6 \ln \frac{6}{5 + v_1/v_s} + \ln \frac{v_1}{v_s} \tag{2.13}$$

In either case γ^0_a has a maximum value of unity when $v_1 = v_s$, and is significantly less than unity if v_s/v_1 is large, as is the case in GC. Thus, athermal effects are always expected to lead to *negative* deviations from Raoult's law, and hence to retention volumes *larger* than expected on the grounds of vapour pressure. Values of γ^0 for hydrocarbons on paraffinic stationary phases (where γ^0_t is close to unity) are found to be less than unity, and are also found to decrease with increasing solvent molecular weight, in agreement with equations (2.12) and (2.13). Also, it is interesting to note that $\gamma^0 (= \gamma^0_t \gamma^0_a)$ for aromatic compounds is less than unity in high molecular weight paraffin solvents but is greater than unity in decane (see Figure 2.1). This is presumably because the solute–solute and solvent–solvent interactions are sufficiently different to make γ^0_t significantly greater than unity for all paraffin solvents while the product $\gamma^0_t \gamma^0_a$ only becomes less than unity at a sufficiently large value of v_s/v_1. Hence, γ^0_a can sometimes be an important determining factor in solvent selectivity.

For sufficiently large values of v_s/v_1, however, both equations (2.12) and (2.13) become

$$\ln \gamma^0_a \approx 1 + \ln \frac{v_1}{v_s}$$

Thus, for two solutes of molar volumes v_1 and v_2, the ratio of athermal activity coefficients in a given solvent is simply

$$\gamma^0{}_{a(1)}/\gamma^0{}_{a(2)} = v_1/v_2 \qquad (2.14)$$

Thus the ratio is unaffected by the nature of the liquid phase provided that v_s/v_1 is large (greater than about 6). If, under these conditions, components

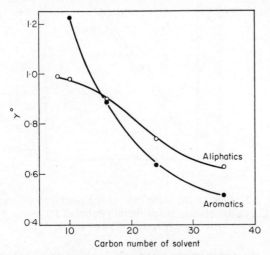

FIGURE 2.1   Plot of the average activity coefficient, γ^0, for aliphatic and for aromatic solutes against the number of carbon atoms in n-alkane solvents. Values of γ^0 less than 1 are probably due to molar volume effects. (From ref. 20)

1 and 2 have identical vapour pressures and identical values of $\gamma^0{}_t$, there will be a *tendency for the smaller molecules to be retained longer than the larger molecules* since, from equations (2.11) and (2.14),

$$r_{12} = P_2 v_2/P_1 v_1 \qquad (2.15)$$

This result possibly accounts for the fact that alicyclic hydrocarbons have larger retention volumes on squalane than do the aliphatic hydrocarbons of the same boiling points (Figure 2.2). Since, for hydrocarbons, equal boiling points imply approximately equal vapour pressures, it follows from equation (2.15) that the observed difference in the logarithms of retention volumes of such aliphatic and alicyclic materials of about 0·07 log unit corresponds to a molar volume ratio of about 1·2. This result is reasonable, since the molar volume of methylcyclohexane (B.P. 100·9 °C) is 128 cc while that of n-heptane (B.P. 98·4 °C) is 148 cc.

It is clear from equation (2.15) that, since P and v are fixed under given conditions, athermal effects, as measured by $\gamma^0{}_a$, can be ignored as sources of variation in *relative* retention of solutes on different stationary phases,

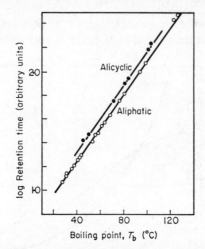

FIGURE 2.2   Plot of the logarithm of the retention on squalane at 43 °C against boiling point for aliphatic and alicyclic hydro-carbons. (From p. 334 of ref. 1)

provided that the assumptions upon which equation (2.15) are based are obeyed, in particular, that the ratio of the partial molar volume of the solvent to that of the solute is greater than about 6, as is usually the case. Therefore, from the point of view of identifying compounds from relative retention behaviour on several columns involving liquid phases of high molecular weight, $\gamma^0{}_t$ is likely to be the only significant differentiating factor [equation (2.11)].

2.2b   Excess enthalpy of mixing

The thermal activity coefficient, $\gamma^0{}_t$, is given by

$$\ln \gamma^0{}_t = \Delta H^0{}_{excess}/RT \qquad (2.16)$$

The excess enthalpy of solution $\Delta H^0{}_{excess}$, in turn, depends upon the relative values of the attractive energies of interaction (E) between solvent and solvent (ss), solute and solute (11), and solute and solvent (1s) molecules. With the exception of the special class of regular solutions (see Section 2.2d), no exact theory is available for predicting $\Delta H^0{}_{excess}$ for a given system.

However, it is generally found that if

$$(E_{ss} + E_{11}) < 2E_{1s}$$

then $\Delta H^0_{excess} < 0$, and therefore $\gamma^0_t < 1$, while if

$$(E_{ss} + E_{11}) > 2E_{1s}$$

$\Delta H^0_{excess} > 0$, and $\gamma^0_t > 1$. Positive values of ΔH^0_{excess}, corresponding to positive deviations from Raoult's law, are the more common. Values of γ^0_t are therefore usually greater than unity, in contrast to athermal activity coefficients, γ^0_a, which, in GC, are invariably less than unity, as discussed above. The magnitudes of the various intermolecular forces at play determine the exact value of γ^0_t and hence the difference in *relative* retention of solutes on different stationary phases.

2.2c Physical interactions in solution

Molecules can be classified into three, very broad categories.

(1) *Polar molecules*, which possess permanent dipole moments.

(2) *Polarisable molecules*, which have mobile electron clouds (π-bonds or lone pairs) which can be polarised by a neighbouring dipole, but do not possess permanent dipoles.

(3) *Non-polar molecules*, which do not possess permanent dipole moments and cannot be significantly polarised.

Forces between molecules can, in turn, be divided into three groups.

(1) *Dispersion forces* result from interactions between instantaneous dipoles produced by the oscillations of nuclei and electron clouds. They alone are responsible for the total cohesive energy in non-polar and polarisable liquids, and even for most of the energy in polar liquids. Solute–solvent interactions in non-polar–non-polar, non-polar–polarisable, non-polar–polar, and polarisable–polarisable types of solution consist entirely of dispersion forces.

(2) *Induction forces* arise from interactions between permanent dipoles and dipoles induced in polarisable molecules. Such forces are much smaller than either dispersion or orientation forces (see below), but can nevertheless play a significant part in solvent selectivity.

(3) *Orientation forces* arise from interactions between permanent dipoles, and therefore, at low solute concentrations, are found only with polar solvents. Their strength depends on the magnitudes and relative positions of the dipoles.

A fourth group of forces, rarely significant in most GC work, but potentially very important, includes chemical forces such as those involved in hydrogen-bonding and complex-formation. 'Specific' chemical forces, in contrast to physical forces, can cause negative deviations from Raoult's law, and hence give rise to values of γ^0_t less than unity. There is no satisfactory theory of solutions which allows γ^0_t to be calculated for all types of solution. However, for the special class of 'regular' solutions, which involve only dispersion forces and have no excess entropy of mixing ($\gamma^0_a = 1$), there is a well developed theory.

2.2d   Regular-solution theory

The theory of regular solutions is based on the concept of internal pressure in a liquid. This pressure, Π, manifests itself by the change in internal energy of a liquid with changes in volume, V, and is defined as $(\partial E/\partial V)_T$. The internal energy of a solution of volume V is therefore ΠV. The internal pressure of a mixture of N_1 moles of solute and N_s moles of solvent in a total volume V can be broken down into three terms, equivalent to three sources of internal energy: $a_{11}(N_1/V)^2$ for solute–solute interactions; $a_{ss}(N_s/V)^2$ for solvent–solvent interactions; and $2a_{1s}N_1N_s/V^2$ for solute–solvent interactions, where a_{11}, a_{ss}, and a_{1s} are constants. Furthermore, if we assume no volume change on mixing,

$$V = N_1 v_1 + N_s v_s$$

where v_1 and v_s are the molar volumes of the pure solute and solvent, respectively. Hence, the total internal energy of the solution is

$$\Pi V = (a_{11} N_1^2 + 2a_{1s} N_1 N_s + a_{ss} N_s^2)/(N_1 v_1 + N_s v_s)$$

while the total internal energy of the separated components is

$$a_{11} N_1/v_1 + a_{ss} N_s/v_s$$

Thus, per mole of solute, the change in internal energy due to mixing is

$$\Delta E^0_{\text{excess}} = \frac{(v_1 N_s v_s)}{(N_1 v_1 + N_s v_s)} \left(\frac{a_{11}}{v_1^2} - \frac{2a_{1s}}{v_1 v_s} + \frac{a_{ss}}{v_s^2} \right)$$

Assuming $a_{1s} = \sqrt{(a_{11} a_{ss})}$ (this is a key assumption, expected to be strictly true only for dispersion forces between spherical molecules), and making the approximation that, under chromatographic conditions,

$$N_1 v_1 + N_s v_s \approx N_s v_s$$

we obtain

$$\Delta E^0_{\text{excess}} = v_1 [\sqrt{(a_{11})}/v_1 - \sqrt{(a_{ss})}/v_s]^2 \tag{2.17}$$

Also,

$$\Delta E^0_{\text{excess}} \approx \Delta H^0_{\text{excess}} = RT\ln\gamma^0_t$$

Using the terminology of Hildebrand,[6] we finally obtain

$$RT\ln\gamma^0_t = v_1(\delta_1 - \delta_s)^2 \qquad (2.18)$$

(Note that, according to this equation, regular solution theory implies that γ^0_t is always greater than unity, i.e. that deviations from Raoult's law are always positive.) δ is called the 'solubility parameter' and is equal to the square root of the internal pressure, Π, of the pure liquid. Π manifests itself as the force which resists the pulling apart of the molecules of a liquid. Hence, since the internal pressure of a vapour is near-zero, the molar energy of vaporisation can be equated with the internal energy of a liquid, Πv. Thus,

$$\Pi = \delta^2 \approx \Delta E^0_{\text{vapn}}/v \qquad (2.19)$$

This approximate equation affords a simple means of calculating values for any solvent and solute, and hence, via equation (2.18), the value of γ^0_t for such a system. This, in turn, enables K and V_g to be calculated from equations (2.7), (2.8), and (2.10) for any regular solution since γ^0_a is then, by definition, equal to unity. δ values for a variety of materials are tabulated in ref. 9.

Relative retention and regular-solution behaviour

According to equation (2.11) the relative retention of two solutes, both of which form regular solutions with the solvent, is given by

$$r_{12} = \gamma^0_{t(2)} P_2/\gamma^0_{t(1)} P_1 \qquad (2.20)$$

Since

$$-RT\ln P = \Delta H^0_{\text{vapn}} - T\Delta S^0_{\text{vapn}}$$

and

$$\Delta H^0_{\text{vapn}} = \Delta E^0_{\text{vapn}} + RT = \delta^2 v + RT$$

we obtain, from equations (2.18) and (2.20),

$$RT\ln r_{12} = T(\Delta S^0_{\text{vapn(2)}} - \Delta S^0_{\text{vapn(1)}}) + \delta_s^2(v_2 - v_1) - 2\delta_s(v_2\delta_2 - v_1\delta_1) \qquad (2.21)$$

For the special case of two solutes which have identical boiling points, identical energies of vaporisation, and hence identical values of ΔS^0_{vapn}, equation (2.21) then reduces to

$$RT\ln r_{12} = \delta_s^2(v_2 - v_1) - 2\delta_s(v_2^{\frac{1}{2}} - v_1^{\frac{1}{2}})(\Delta E^0_{\text{vapn}})^{\frac{1}{2}} \qquad (2.22)$$

This equation demonstrates the importance of molar volume in determining relative retention for solutes forming regular solutions, since, even though solutes may have identical boiling points and energies of vaporisation, they only have the same retention volumes if they have the same molar volumes. If $v_1 \neq v_2$, the greatest relative retention will be obtained on a liquid phase with a large solubility parameter, δ_s, the smaller molecules then being retained most strongly [cf. $\gamma^0{}_a$ effects; equation (2.15)], although the order of elution may be reversed at low values of δ_s.

Benzene ($T_b = 80 \cdot 1 \,^\circ$C) and cyclohexane ($T_b = 80 \cdot 9 \,^\circ$C) have very similar heats of vaporisation, and therefore equation (2.22) may be expected to apply, albeit approximately. Any differences in retention for these two substances in regular solutions can be attributed simply to the differences in molar volumes. Calculated values[7] for relative retentions of benzene and cyclohexane on isoquinoline and 4-methylquinoline are in good agreement with calculation if, in order to validate equation (2.22), it is assumed that $\Delta S^0{}_{\mathrm{vapn}(1)} = \Delta S^0{}_{\mathrm{vapn}(2)}$. Such agreement may well be fortuitous, since there is a difference of 200 cal mole^{-1} in the heats of vaporisation of the two solutes, which would correspond to an incorrect vapour pressure ratio of cyclohexane to benzene of about $1 \cdot 4$ at 20 $^\circ$C if entropies of vaporisation were indeed equal, as assumed.

Martire[8] made no such assumption for cyclohexane and benzene on dimethylformamide, but obtained reasonable agreement with experimentally measured *absolute* retentions only after making corrections to equation (2.18), which became

$$RT \ln \gamma^0 = v_1[(\delta_1 - \delta_s)^2 + (\omega_1 - \omega_s)^2 - c_s(v_1/\delta_1)] \qquad (2.23)$$

where c_s is a semi-empirical constant, characteristic of the stationary phase, and $(\omega_1 - \omega_s)^2$ makes allowance for polar interactions. ω is given by

$$\omega \approx \mu/\sqrt{(2 \cdot 31 v^3 kT)}$$

where μ is the dipole moment of the molecule and k is Boltzmann's constant. Corrections of this kind are obviously of limited utility, and serve to emphasise the current difficulties of theories of solution.

It is impracticable to use equations such as (2.18) and (2.21) for the accurate prediction of retention data, even for strictly regular solutions. It is known, for example, that the values of δ for n-alkanes which are applicable in solution, although fairly constant, are about 1 cal$^{\frac{1}{2}}$ cm$^{-\frac{3}{2}}$ larger than those calculated from heats of vaporisation and molar volumes. Furthermore, since the estimation of δ_s values depends on the latent heat of vaporisation of the solvent, they can only be calculated for the most volatile stationary phases of well defined composition, and even then use must normally be made of approximate relations between boiling points and heats of vaporisation. Small errors in estimated heats of vaporisation of 100 cal mole^{-1} can give rise to large final errors in calculated retention ratios. Finally, the theory of regular solutions ignores the athermal or entropy term, as well as solution forces arising from polar interactions, although modifications of equation (2.18) to take account of these effects have been proposed, such as equation (2.23).

Nevertheless, the solubility parameter of a solvent provides a useful guide to its selectivity,[9] even when the presence of polar forces in solution causes the geometric mean assumption, $a_{1s} = \sqrt{(a_{11} a_{ss})}$ to break down, and hence invalidates equation (2.18). Thus, in general it is found that $\gamma^0{}_t$ increases (i.e. retention decreases) as the difference between δ and δ_s increases, and that retention is greatest when $\delta = \delta_s$, even for polar solvents. This conclusion is an algebraic statement of the aphorism 'like dissolves like'.

2.2e Polar solvents

Since a non-polar solute experiences only dispersion forces, even in a highly polar solvent, it might be expected at first sight that the retention volume of such a solute on any phase will be little different from that predicted on the basis of vapour pressure. However, in the process of dissolving in a polar solvent, a considerable amount of energy has to be expended in order to separate solvent molecules and disrupt orientation forces. Thus, in general, as the polarity of the solvent increases, $\Delta H^0{}_{\text{excess}}$, and hence $\gamma^0{}_t$, for a non-polar solute will also increase. It follows that the retention of non-polar solutes on polar solvents will be smaller than on non-polar solvents under similar conditions. Increasing attention is being paid to the development of polar liquid phases which consist of molecules with considerable non-polar portions, so that the retention volumes of non-polar solutes on these liquids are similar to those observed on non-polar liquids such as squalane, while polar solutes are preferentially retained.[10]

A polarisable solute such as an olefin experiences induction forces in polar solvents which partially compensate for the disruption of orientation forces in the solvent resulting from its solution. Thus, on a non-polar stationary phase (e.g. squalane) an olefin is eluted at essentially the same time as a paraffin with the same boiling point, while on a 'slightly polar' liquid such as polyethylene glycol the relative retention of olefin to paraffin increases to around 1·2. The corresponding ratios on the 'very polar' phases β,β'-oxydipropionitrile and dimethylsulpholan are even larger, 1·9 and 2·2 respectively. Thus, the retention volumes of polarisable molecules on polar columns are very much larger than those expected from a consideration of their boiling points and the corresponding retention volumes of the paraffins. In thermodynamic terms, this result is obtained because in a polar solvent $\gamma^0(\text{paraffin}) > \gamma^0(\text{olefin}) > 1$; in physical terms, it is due to the polarisability of the olefin allowing it to interact more strongly with the solvent.

The relative retention of polar solutes and their parent paraffins is also greater on polar solvents than on non-polar solvents. However, in this case a moderately polar solvent can be considered to be non-selective towards polar solutes ($\gamma^0 \approx 1$), since it may well separate them strictly according to their vapour pressures. This result reflects the similarity of interactions in solute and solvent.

2.2f   Non-polar solvents

Solution of a polar solute in a non-polar solvent eliminates the dipole–dipole orientation forces present in the liquid solute, thereby reducing its retention volume below that expected from consideration of vapour pressure or boiling point. In thermodynamic terms, these solutions exhibit positive deviations from Raoult's law ($\gamma^0 > 1$), because (cf. p. 15)

$$E_{1s} \approx E_{ss} \ll E_{11}$$

and hence

$$E_{1s} \ll \frac{E_{11} + E_{ss}}{2}$$

Thus, on squalane, an ester (polar) is eluted at about the same time as a paraffin of the same molecular weight, even though its boiling point is considerably higher. Hence, the term 'boiling point column', often applied to non-polar solvents, really applies only to the separation of non-polar and polarisable solutes, and *not* to polar solutes. Although facile terminology is to be avoided, a better term if needed might be 'molecular weight column'. Figure 2.3 illustrates a rough linear correlation between $\log_{10} V_g$ and molecular weight for a variety of compounds of molecular weight between 30 and 130. Such columns are obviously useful in qualitative analysis, since the retention volume of a solute sets limits for its molecular weight. However, a polar column alone is less useful, since compounds having the same retention volumes may have widely different molecular weights.

2.3   COLUMN SELECTIVITY IN PRACTICE

The absence of a firm theory of solutions, as well as the lack of detailed data for γ^0 for GC systems, means that in practice column selectivity must often be considered from a qualitative, semi-empirical point of view. The idea of one column being 'more polar' than another is commonly used in GC when what is frequently meant is that the two columns offer substantially different *selectivities* for a particular separation (see p. 11).

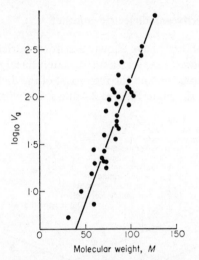

FIGURE 2.3 Plot of the logarithm of the specific retention volume on squalane at 80 °C against molecular weight for a variety of compound types, including hydrocarbons, alcohols, and esters. The line is drawn through the points for n-alkanes. (From ref. 11)

It should be noted that although a column 'polarity' scale is feasible, at least in principle, a column selectivity scale is not, since any particular column will be non-selective to an eluate of similar 'polarity' to itself, and selective to eluates with higher or lower polarities. For an unknown eluate the vapour pressure will not generally be known, so that it will not be possible to decide *a priori* whether a particular column is behaving selectively or not.

This consideration serves to stress a fact which is often obscured in gas chromatographic texts, namely, that the solvent knows nothing about the vapour pressure or boiling point of the solute. Only solvent–solvent and solvent–solute interactions determine the retention behaviour at infinite dilution. Solute–solute interactions are only important insofar as they act as a guide to the magnitude of solvent–solute forces, such as in regular solution theory.

Some methods of defining 'polarity' for liquid phases will now be considered. Methods of column classification have been usefully discussed by Littlewood[11] and by Rohrschneider,[12] who both conclude that, although there are certain general trends, it is not possible to predict accurate retention behaviour on a particular stationary phase from any single 'polarity' parameter.

2.3a   Column polarity and dielectric constant

Anvaer and co-workers[13] have shown that for a variety of hydrocarbons and oxygenated solutes there is an approximately linear correlation between the free energy of solution and the reciprocal of the dielectric constant ε of the solvent. This is shown in Figures 2.4 and 2.5 in which $\log_{10} K$ is plotted

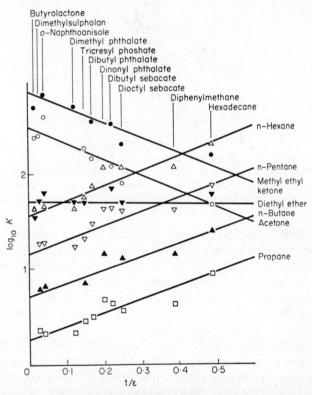

FIGURE 2.4   Plot of the logarithm of the partition coefficient at 40 °C for various solutes against the reciprocal of the dielectric constant of the stationary phase. (From ref. 13)

against $1/\varepsilon$ for values of ε ranging from 2 for hexadecane to 32 for dimethylsulpholan and 37 for butyrolactone. The dielectric constant of a solvent is a more valuable indicator of its 'polarity' than is its dipole moment. For example, the dipole moment of a symmetrical molecule may be zero, even though its individual bonds are of a highly polar nature and its dielectric

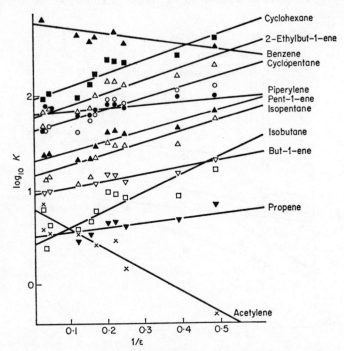

FIGURE 2.5 Plot similar to Figure 2.4, for various hydrocarbons. (From ref. 13)

constant correspondingly high. Moreover, it is known that the dipole moment of a homologous series such as the alcohols is almost constant, whereas the dielectric constant falls steadily with increasing chain length.

Anvaer and co-workers argue that the energy of solution, and hence $\log V_g$, of a polar solute is expected to increase with increasing solvent dielectric constant, since the net gain in energy on dissolving a dipole in a dielectric is given by

$$\frac{2}{9}\frac{\mu^2}{r^2}\left(1-\frac{1}{\varepsilon}\right)$$

where μ is the dipole moment and r is the radius of the solute molecule, and ε is the dielectric constant of the solvent. However, the energy of solution of a non-polar solute should *decrease* with increasing dielectric constant, since progressively more work is expended in displacing the solvent

B

molecules than is recovered by solute–solvent interactions. These predictions are in general borne out by Figures 2.4 and 2.5, although in this sense benzene and acetylene appear to be 'polar' and olefins to be slightly 'non-polar'.

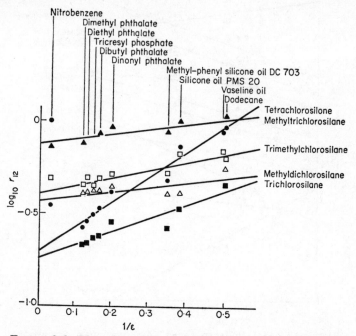

FIGURE 2.6   Plot of the logarithm of the retention volumes of various chlorosilanes, relative to dimethyldichlorosilane, against the reciprocal of the dielectric constant of the stationary phase.
(From ref. 13)

Figure 2.6 shows similar data, this time using relative retentions, for some chlorosilane solutes, and again a reasonable correlation between retention behaviour and ε is observed for values of ε between 2 and 8, but nitrobenzene ($\varepsilon = 35$) behaves as if it had a dielectric constant of only 6, especially towards tetrachlorosilane.

Thus, although there must remain some doubt about the detailed validity of this approach, it does suggest that the solvent dielectric constant may provide the basis for a useful general scale of column 'polarity'.

2.3b   Rohrschneider plots

Rohrschneider[14] has proposed a graphical method for determining the polarity of a liquid phase, one form of which is illustrated schematically in

Figure 2.7. Each vertical line represents the chromatogram, obtained on a single liquid phase for the same solutes, with the retention increasing logarithmically from the bottom. Peaks for a standard solute, represented by filled circles, are lined up on a horizontal line corresponding to a relative

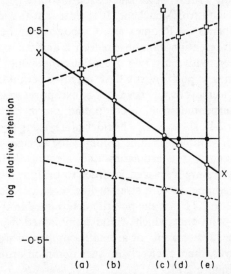

FIGURE 2.7   Schematic Rohrschneider plot for the logarithms of relative retentions on five columns, (a) to (e). The standard compound for relative retention measurements is shown ●, while the reference compound shown ○ is arranged on the straight line ×—×. The compound △ has no specific interactions on any column, but compound □ may have specific interactions on column (c)

retention of unity on all columns. The vertical chromatogram lines are then ranged horizontally until the peaks corresponding to the reference compound (open circles) form a straight diagonal line. When this is done, it may be observed that the peaks for other compounds also lie approximately on straight lines, as shown schematically by the triangles in Figure 2.7. The horizontal separation of the chromatogram lines then provides a direct scale on which to measure polarities, with the two extreme phases, corresponding to polarities of, say, 0 and 1, providing fixed points. If a particular peak on a particular column is out of line, as indicated by the square on column (c) of Figure 2.7, it is possible that specific interactions are involved. If this particular solute were to be chosen as a standard or reference, then all other peaks on that stationary phase would be out of line.

Figure 2.8 shows a Rohrschneider plot for Apiezon-L (paraffinic), Reoplex 100 (polyester), Carbowax 4000 (polyglycol), and β,β'-oxydipropionitrile constructed for various acyclic (a), monocyclic (b), and bicyclic (c) terpenes, with limonene taken as the standard compound.[15] The chromatogram lines for Reoplex and Carbowax have been placed with the aid of retention data for β-fenchene. It is clear that the relative retentions of the other terpenes on Reoplex and Carbowax can be predicted fairly well simply from retention data on Apiezon-L and β,β'-oxydipropionitrile columns. Further, with only two exceptions the slopes of the lines for particular terpenes as positive, small, or negative permit correct classification as acyclic, monocyclic, or bicyclic. The exception seen in Figure 2.8(b) is in fact a non-terpenoid aromatic compound, p-cymene.

Figures 2.9 to 2.12 show Rohrschneider plots, using specific rather than relative retention volumes on nine columns for straight-chain and cyclic alkanes and alkenes, for aromatics and alcohols.[11] In all cases the chromatogram lines are displaced in order to bring the specific retention volumes for n-heptane on to a straight line of unit slope. It can be seen from Figures 2.9 to 2.11 that the polarity determined in this way coincides reasonably well with that which would be obtained if another paraffin, olefin, naphthene, cycloalkene, or aromatic compound were to be used as standard. However, there is clearly no such common order of polarity for the alcohols of Figure 2.12.

In Figure 2.10 it is interesting to note the cross-over as column polarity is increased with alkenes being eluted after the corresponding alkane, instead of before.

FIGURE 2.8   Rohrschneider plots of the logarithms of retentions of various terpenes relative to limonene on columns of Apiezon-L (A), Reoplex 100 (B), Carbowax 4000 (C), and β,β'-oxydipropionitrile (D). The positions of B and C have been chosen so that the broken line (for β-fenchene) is straight. (a) Acyclic; (b) monocyclic; (c) bicyclic terpenes. (Data from ref. 15)

1, *trans,trans*-Allo-ocimene	12, α-Phellandrene
2, *trans,cis*-Allo-ocimene	13, β-Fenchene
3, Ocimene Y	14, Sabinene
4, Ocimene X	15, Δ^3-Carene
5, Myrcene	16, β-Pinene
6, β-Fenchene	17, α-Fenchene
7, p-Cymene	18, β-Fenchene
8, Terpinolene	19, α-Pinene
9, γ-Terpinene	20, Tricyclene
10, β-Phellandrene	21, Santene
11, α-Terpinene	

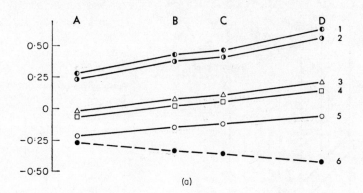

(a)

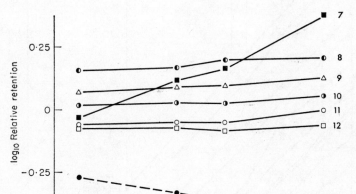

(b)

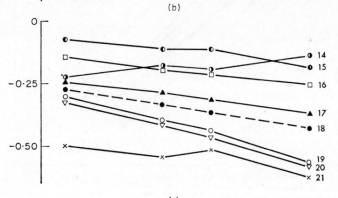

(c)

FIGURE 2.8

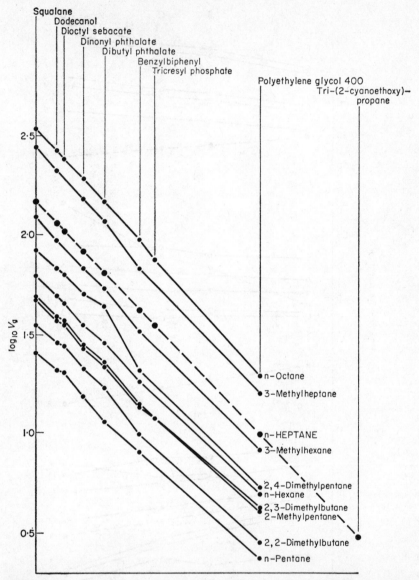

FIGURE 2.9   Rohrschneider plot of the logarithm of the specific retention volumes of alkanes on nine stationary liquids at 80 °C. The chromatogram lines are arranged so that the points for n-heptane form a straight line of unit slope shown broken. (From ref. 11)

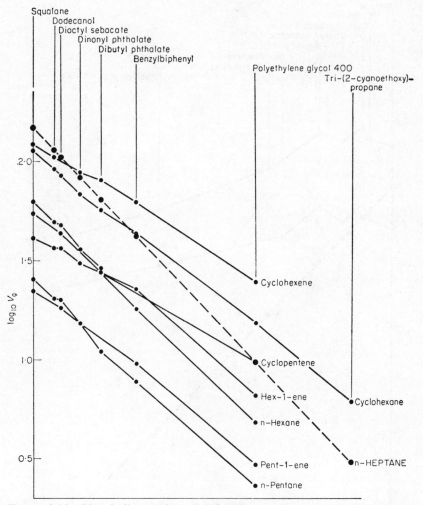

FIGURE 2.10   Plot similar to Figure 2.9, for alkanes, alkenes, cycloalkanes, and
cycloalkenes. (From ref. 11)

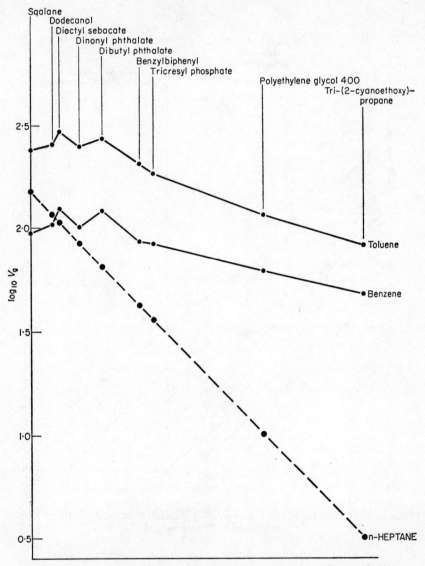

FIGURE 2.11   Plot similar to Figure 2.9, for aromatics. (From ref. 11)

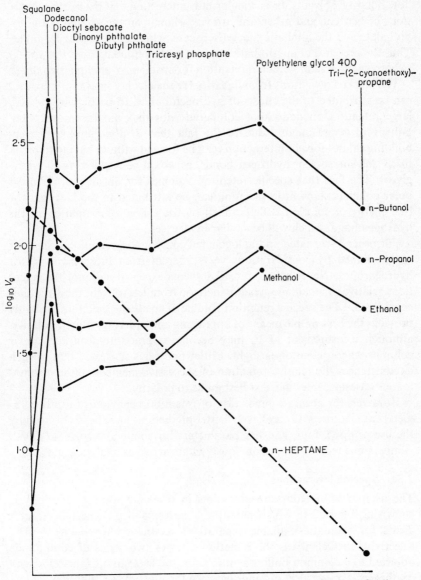

FIGURE 2.12 Plot similar to Figure 2.9, for alcohols. (From ref. 11)

In Figure 2.11 the lines are more crooked than in Figures 2.9 or 2.10 for 'low polarity' solvents. In particular, it is interesting that the relative retentions of benzene and n-heptane are very similar on the aromatic dinonyl phthalate and the aliphatic dioctyl sebacate, which have similar formulae. Thus the presence of an aromatic nucleus in the phthalate does not appear to correspond to a preferential retention for aromatic over aliphatic solutes.

The marked deviations from linearity for the alcohol lines in Figure 2.12 can be attributed to the effects of hydrogen bonds. In particular, the very large retention on dodecanol compared with that on the 'more polar' polyglycol is presumably due to the fact that in dodecanol hydrogen-bonding solute–solvent interactions can be formed without having to break many solvent–solvent hydrogen bonds, as would be required in the polyglycol. The fact that specific retention volumes for an alcohol on other polar columns show little variation can be attributed to the fact that the free energy of solution is determined by the energy of formation of the hydrogen bond, which will be similar in all cases.

For other oxygenated and halogenated compounds, retention behaviour is determined by considerations very different from those involved for hydrocarbons. Nevertheless, for all solvents except dodecanol the retentions relative to an alkane steadily increase from left to right for all solutes considered.[11] The specific retention volume of an alkane therefore probably provides the best simple means of appraising and ordering column polarity, although measurement of V_g may become impracticable on very polar columns where retention is slight. Littlewood[11] has suggested that in such circumstances the relative retention of successive members of any homologous series provides the next best guide to polarity.

However, for accurate prediction of retention behaviour, several parameters are required. Progressively better predictions have been obtained by the use of three, four, and five parameters for solute and solvent.[16] Such cumbersome techniques are obviously unattractive as analytical aids.

2.3c   Specific interactions

The usefulness of Rohrschneider plots in revealing solvent specificity is shown in Figure 2.13 which plots the logarithms of the selectivity coefficients for aromatic–n-alkane separations as the reference line.[17] (The selectivity coefficient[18] is the retention ratio of two series of compounds adjusted to a common boiling point. Thus, the logarithm of the selectivity coefficient is the vertical distance between the plots of log retention against boiling point.) It can be seen that the selectivity coefficients for n-alkene–n-alkane separation are in general agreement with those for the aromatic–n-alkane systems, with the notable exception of the glycol–silver nitrate column, which, as expected for a complexing agent, shows a marked specific retention for alkenes.

Another commonly quoted 'specific' stationary phase is the 1:1 fluorene–picric acid addition compound, which certainly retards aromatics

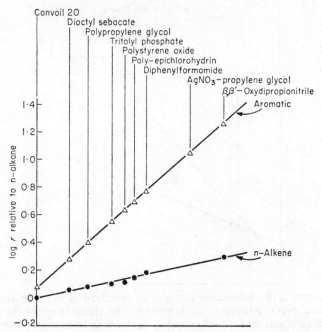

FIGURE 2.13 Rohrschneider plot of retention of an aromatic hydrocarbon and an n-alkene relative to an n-alkane, all at a common boiling point of 100 °C. The chromatogram lines are arranged so that the aromatic points fall on a straight line. There is a significant deviation for the alkene on the AgNO₃–glycol column. (From ref. 17)

more strongly than, for example, olefins. However, as shown in Figure 2.14, the Rohrschneider plot simply reveals fluorene–picric acid as a moderately polar stationary phase, without any *specific* retention for benzene.[17] Thus, the retention volume of benzene on fluorene–picric acid is that expected from the retention of cyclohexene on that phase, together with the retentions of benzene and cyclohexene on the non-specific paraffinic and polyglycol-containing phases, all retentions being measured relative to that of cyclohexane.

Figure 2.15 shows that the highly 'polar' phase 1,2,3-tri-(2-cyanoethoxy)propane, although showing great selectivity between paraffins and aromatics, is also without specificity, in agreement with Figure 2.11, while the phases di-n-butyl tetrachlorophthalate and 2,4,6-trinitrophenetole do *specifically* retain benzene. It is interesting to note that on both di-n-butyl

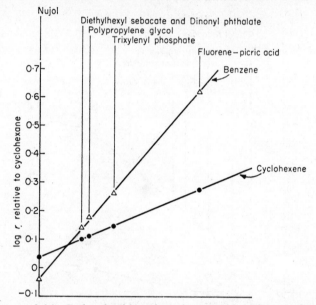

FIGURE 2.14   Rohrschneider plot of retention of benzene and cyclohexene relative to cyclohexane. The chromatogram lines are arranged so that the cyclohexene points fall on a straight line; the benzene points are then also almost linear, even for fluorene–picric acid. (From ref. 17)

tetrachlorophthalate and 2,4,6-trinitrophenetole the usual *p*- and *m*-xylene emergence order observed on other phases, including fluorene–picric acid, is reversed.

Examples of specific interactions of this kind, between solute and liquid phases, are extremely rare. Apart from those mentioned above and other halogenophthalate–aromatic systems, there are very few other specific stationary phases, and these usually involve cationic complexes. This is an area ripe for exploitation.[19]

2.3d   Conclusions

If no facilities are available for other GC identification techniques, it may be worthwhile to attempt to confirm a tentative identification by the use of a variety of columns of intermediate 'polarity' but as different as possible in chemical type. The phenyl- and vinyl-substituted silicone phases and the hydrogen-bonded polyglycols are useful for such work.

Broadly speaking, in the absence of specific interactions, the 'polarity' of the stationary phase provides only one degree of freedom for purposes of identification. This is particularly true for hydrocarbons. Thus, the quickest

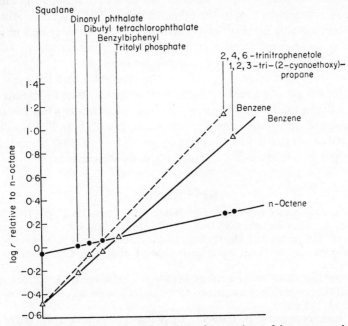

FIGURE 2.15   Rohrschneider plot of retention of benzene and n-octene relative to n-octane. The chromatogram lines are arranged so that the n-octene points fall on a straight line. The benzene points on dibutyl tetrachlorophthalate and on trinitro-phenetole suggest specific retention. (From ref. 17)

way to obtain most useful qualitative information about an unknown compound is to use two columns differing as widely as possible in 'polarity'. There is little difficulty in obtaining non-polar stationary phases of high boiling point, but the more common 'polar' phases can only be used at moderate temperature. For example, β,β'-oxydipropionitrile has an effective upper limit of 70 °C. Tri(cyanoethoxy)propane is better in this respect since it has a very similar polarity and can be used up to about 150 °C. Thus, a paraffinic stationary phase such as squalane or Apiezon, together with one of these highly polar phases, provides the basic materials for identification by means of retention. (It should be noted that there are often subtle differences between retention behaviour on squalane and on a methyl siloxane polymer.) If the solutes and solvents involved roughly obey a Rohrschneider plot, the use of columns of intermediate polarity can be of little more than confirmatory value except when peaks in a very complex mixture are to be analysed. The use of many columns of gradually increasing polarity may then help to track the movement of peaks from one side of the Rohrschneider plot to the other, so unravelling what may otherwise be an unresolvable tangle of peaks. However, it is probably best in such a situation to attempt some preliminary class separation, or to transfer individual peak complexes from one column to another.

After the preliminary diagnosis on the non-polar and the very polar columns, the use of a specific reagent such as silver nitrate for olefins or an alkyl halogenophthalate for aromatics may be indicated, but the use of such columns in routine identification work is unlikely to be helpful. The potentialities of gas–solid adsorption columns should always be considered, since there is then much more scope for specific interactions.

REFERENCES

1. H. Purnell, *Gas Chromatography*, Wiley, New York (1962), p. 67.
2. Gas Chromatographic Section of G.A.M.S., 'Values of the correction factor *j* of James and Martin used in gas chromatography, for values of the ratio of inlet to outlet pressure between 1 and 3', *J. Chromatog.*, **2**, D33 (1959).
3. American Society for Testing and Materials, Committee E-19, 'Recommended practice for gas chromatography terms and relationships', A.S.T.M. designation E 355–68.
4. S. H. Langer and J. H. Purnell, 'A gas–liquid chromatographic study of the thermodynamics of solution of some aromatic compounds', *J. Phys. Chem.*, **67**, 263 (1963).
5. D. E. Martire, 'Thermodynamics of solutions as studied through gas–liquid chromatography', *Fourth International Symposium on Gas Chromatography* (ed. L. Fowler), Academic Press, New York (1963), p. 33.
6. J. H. Hildebrand and R. L. Scott, *Regular Solutions*, Prentice-Hall, Englewood Cliffs, N.J. (1962).
7. L. Rohrschneider, 'Gas chromatographic separation in regular solutions', *J. Gas Chromatog.*, **6**, 5 (1968).
8. D. E. Martire, 'Applications of the theory of solutions to the choice of solvent for gas–liquid chromatography', *Analyt. Chem.*, **33**, 1143 (1961).
9. S. H. Langer and R. J. Sheehan, 'Theory and principles for choosing and designing selective stationary phases', in *Progress in Gas Chromatography*, Vol. 6 (ed. J. H. Purnell), Wiley, New York (1968), p. 289.
10. A. B. Littlewood and F. W. Willmott, 'The identity of equivalent mixtures as stationary liquids', *J. Gas Chromatog.*, **5**, 543 (1967).
11. A. B. Littlewood, 'The classification of stationary liquids used in gas chromatography', *J. Gas Chromatog.*, **1** (11), 16 (1963).

12. L. Rohrschneider, 'The polarity of stationary liquid phases in gas chromatography', in *Advances in Chromatography*, Vol. 4 (ed. J. C. Giddings and R. A. Keller), Marcel Dekker, New York (1967), p. 333.

13. B. I. Anvaer, A. A. Zhukhovitskii, I. I. Litovtseva, V. M. Sakharov, and N. M. Turkel'taub, 'On the relationship between retention volume and the dielectric constant of the stationary phase in gas–liquid chromatography', *J. Analyt. Chem.* (*USSR*) (English translation), **19**, 162 (1964).

14. L. Rohrschneider, 'Polarity of the stationary phase in GC', *Z. Analyt. Chem.*, **170**, 256 (1959).

15. M. H. Klouwen and R. Ter Heide, 'A systematic analysis of monoterpene hydrocarbons by gas–liquid chromatography', *J. Chromatog.*, **7**, 297 (1962).

16. L. Rohrschneider, 'A method for the characterisation of gas chromatographic stationary liquids' (in German), *J. Chromatog.*, **22**, 6 (1966). (An English translation by J. Weber, Chemcell Research Laboratories, Edmonton, Alberta, Canada, is available.)

17. A. O. S. Maczek and C. S. G. Phillips, 'Aromatic interactions in gas chromatography: the use of 2,4,6-trinitrophenetole as a column liquid', *J. Chromatog.*, **29**, 15 (1967).

18. E. Bayer, *Gas Chromatography*, Elsevier, Amsterdam (1961), pp. 29 and 184.

19. B. L. Karger, 'New developments in chemical selectivity in gas–liquid chromatography', *Analyt. Chem.*, **39** (8), 24A (1967).

20. A. Kwantes and G. W. A. Rijnders, 'The determination of activity coefficients at infinite dilution by gas–liquid chromatography', in *Gas Chromatography*, 1958 (ed. D. H. Desty), Butterworths, London (1958), p. 125.

CHAPTER THREE

IDENTIFICATION BASED ON RETENTION

The most common method of identification used in GC is undoubtedly that of 'identification from retention'. In the early days of GC the simplicity, low cost, and wide applicability of the technique made it an obvious first choice and probably did much to delay the development of more effective and reliable methods. Retention methods are most useful when the nature of the unknown eluate is fairly well established and what is required is an exact identity, or the confirmation of a tentative assignment. On the other hand, identification of a completely uncharacterised sample from retention data alone is often not very reliable and may be virtually impossible.

Any problem of identification is generally divisible into three, broad stages. First, some overall guidance as to the identity of the sample is sought. Secondly, substantiation of this is sought, generally through appropriate experimentation. Thirdly, a tentative assignment is confirmed, either by carrying out some critical experiment or by a process of elimination involving exhaustive consideration of all other possibilities.

When attempting identification of eluates from retention data, guidance may be given (a) by comparing published retention data for known substances with experimental data for the unknown in a search for possible retention coincidences, (b) by fitting experimental retention data into established correlation patterns based on carbon number and molecular structure, and (c) by considering retention data in the light of background information about the sample. Substantiation is sought (a) by testing retention coincidence on other, significantly different columns, (b) by exploring retention–structure correlation patterns for other columns, and (c) by seeking repetition of retention coincidence at a variety of temperatures. Confirmation follows only from a detailed and exhaustive examination of the retention data of all possible alternatives such as isomers etc., or by carrying out confirmatory experiments unrelated to retention.

Clearly, any particular problem will not usually involve all these steps.

3.1 RETENTION COINCIDENCE

3.1a Experimental procedure

Retention coincidence is the basis of identification from retention. If standard and unknown eluates have different retention volumes they cannot be the same material, whereas repeated retention coincidence between them on a variety of types of column strongly implies that they are identical. The procedure for narrowing down the number of possible eluates, and for eventually testing retention coincidence, varies with each problem; typically it might be as follows.

(*i*) Background information about the eluate is sought. Its origin is obviously important in limiting its possible identity. All available information is then used to draw up a short list of possible types of compound; without this information identification by retention is virtually impossible.

(*ii*) Retention data for the short-listed materials are collected from the growing body of literature (see, e.g., refs. 1—3). Although it is unlikely that these data can, in practice, be reproduced exactly, they do provide a good guide for the selection of analytical columns. Furthermore, and more importantly, measurement of the retention volume of only one eluate on the actual column in use enables calculation of the factor by which all published retention data for the particular liquid phase must be scaled in order to apply to this column. This fact is illustrated schematically in Figure 3.1, in which logarithms of the published retention volumes for a given column are plotted against experimentally determined values for a similar column involving the same liquid phase. This plot is a straight line with unit slope, so it is completely determined by a single point on it. A plot of published and experimental *relative* retention data should be similar

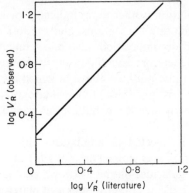

FIGURE 3.1 Schematic log–log plot of experimental values against literature values of adjusted retention data. For a given liquid phase and temperature the plot should be linear, with unit slope

to Figure 3.1, and can be equally useful. (It is, of course, essential that all retention data be properly corrected for column dead-space and apply to the same column temperatures.) It must be emphasised that, in practice, the procedure illustrated in Figure 3.1, or the equivalent calculation, is only approximate, even when temperature, the type of inert support, and the extent of liquid loading are nominally identical.

(*iii*) After any necessary scaling, the published data are then compared with the data for the unknown eluates, and possible coincidences noted. In view of the sometimes poor reproducibility of retention data, it is wise to regard rough agreement between published and experimental data as implying possible coincidence. In this way, it is often possible to say that an unknown material is definitely not one of the vast majority of compounds, but may be one of a few substances. Application of steps (*ii*) and (*iii*) to two columns of substantially different properties may narrow the field even further. Suitable pairs of column solvents are squalane and β,β'-oxydipropionitrile for low-boiling materials, and a methyl siloxane (e.g. SE 30 or OV 1) and a nitrile silicone (e.g. XE 60 or OV 225) for high-boiling materials eluted at temperatures up to 250 °C.

The next step is to compare directly the retention behaviour of the unknown with each of the possibilities, and, unlike the earlier steps in the process, this demands that a sample of each standard is available.

It would be convenient if published retention data could be used with the same confidence as infrared spectra and, to a lesser extent, mass spectra. However, at present this is not possible because of the many factors which affect retention behaviour. Among the most important of these are solid-support effects, and the variation of retention volume with sample size, due to non-linear solution isotherms, as indicated by asymmetrical peaks. Temperature and flow rate fluctuations, and the loss of liquid phase due to column bleed and decomposition, also affect retention volumes, often from day to day. It is therefore essential in any crucial retention coincidence experiment, such as those outlined above in which the unknown has been narrowed down to a few 'possibles', to chromatograph the unknown and standard under as nearly identical conditions as possible. This can be done in at least two ways.

(*a*) The standard is injected in admixture with the sample, and the resulting chromatogram is examined carefully for any new peaks, and for any shoulders or irregularities on the unknown peaks. Observation of any of these phenomena indicates that the standard material is not the same as any of the unknown eluates, whereas an increase in peak-height without corresponding increase in peak-width—conveniently measured at half-height—indicates good retention coincidence.

(*b*) The chromatograms for standard and unknown are obtained separately, but separated by as short an interval of time as possible. The retention volumes of standard and unknown are then compared.

3.1b   Statistical aspects of retention coincidence

In the experiments (a) and (b) outlined above, it is obviously important to know how large the difference between the retention volumes of standard and unknown materials, ΔV_{min}, would have to be before it could be seen as a broadening of the peak in method (a) or as a significant difference of retention times (or volumes) in method (b). Klein and co-workers[4] have discussed model systems akin to method (a). Their technique of measuring peak dispersion by 'probit' analysis of successive fractions is unduly complicated for the present discussion, but essentially similar conclusions for two overlapped peaks can be obtained by measuring the dispersion of a double peak by its width at half-height.

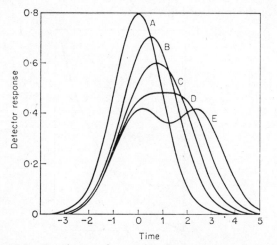

FIGURE 3.2   Envelopes of unresolved gaussian peaks of equal height at separations of (A) 0, (B) 1, (C) 1·5, (D) 2, and (E) 2·5 standard deviations respectively. The time scale is in units of standard deviation (σ) about the retention time of one component

It is convenient to work in terms of the fractional difference, $\Delta V_{min}/V$, in retention volume that can be detected in a double peak, and to characterise a gaussian GC peak by its standard deviation, σ. In chromatographic parlance, σ is related to the number of theoretical plates, n, by the relation $n = (V/\sigma)^2$, so that

$$\frac{\Delta V_{min}}{V} = \frac{\Delta V_{min}}{\sqrt{(n)}\sigma} \tag{3.1}$$

Figure 3.2 shows the envelope of two equal overlapped gaussian peaks with the same standard deviation σ at various separations ΔV. This model applies tolerably well to two equal GC peaks with very similar retention volumes. A gaussian peak has a width at half-height, $W_{\frac{1}{2}}$, of $2\cdot3\sigma$, and this is also the value of $W_{\frac{1}{2}}$ of two exactly coincident peaks ($\Delta V/\sigma = 0$) as shown in Figure 3.2(A). Figure 3.2(B) and (C) show what happens when $\Delta V/\sigma = 1\cdot0$ and $1\cdot5$, the corresponding values of $W_{\frac{1}{2}}$ being $2\cdot7\sigma$ and $3\cdot2\sigma$, although the peak envelopes still appear gaussian to the eye. At a separation of $\Delta V/\sigma = 2\cdot0$, however [Figure 3.2(D)], the peak has a pronounced flat top, and $W_{\frac{1}{2}} = 4\cdot0\sigma$, i.e. almost twice the width of a single peak. It should be noted that the maximum of the peak envelope in Figure 3.2(A)—(D) is shifted $\Delta V/2$ from each component peak maximum. For separations greater than $\Delta V/\sigma = 2\cdot0$, a minimum can be seen between the two peaks [Figure 3.2(E)]. Figure 3.3 shows the percentage increase in $W_{\frac{1}{2}}$ with separation for the peaks of Figure 3.2.

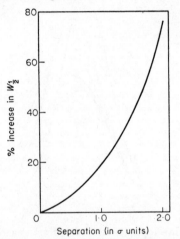

FIGURE 3.3   Effect of difference in retention on the width at half-height, $W_{\frac{1}{2}}$, of the envelope of two gaussian peaks of equal height. Separation is in units of standard deviation, σ

On the reasonable basis that a 10% increase in peak-width would be detectable, it follows that the minimum difference in retention volume detectable from an increased peak-width is given by $\Delta V_{min}/\sigma = 0\cdot7$. From equation (3.1) it follows that for method (a), involving peaks of equal size, $\Delta V_{min}/V \approx 0\cdot7/\sqrt{n}$. With typical values of plate numbers, therefore, retention volumes differing by more than about 1% for packed columns and 0·2% for capillary columns could be distinguished by examination of the width at half-height. The limitation of equal peak size is not very critical since the calculation is intended only as a guide, and peaks can be arranged to be of roughly equal size by adjusting the amount of added standard.

The usefulness of peak-width measurements as a test for overlapped peaks is shown by Figure 3.4, which is a plot of $W_{\frac{1}{2}}$ against adjusted retention time for a variety of alkanes, silanes, and alkylsilanes on silicone oil.[5]

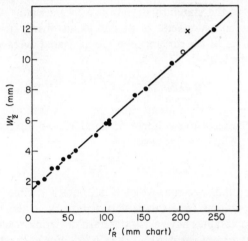

FIGURE 3.4   Use of peak-width to show unresolved components. Plot of peak-width at half-height, $W_{\frac{1}{2}}$, against adjusted retention time, t_R', for various alkanes, silanes, and alkylsilanes. The point × is an overlapped peak, probably containing trisilane (○) impurity. (From ref. 5)

The point × is clearly suspect, and is probably attributable to trisilane impurity in 1,1-dimethyldisilane, both compounds having similar retention times. The point for pure trisilane is shown as ○. The distance of point × from the line corresponds to an increase in $W_{\frac{1}{2}}$ of only about 10%, so that to the eye the peak showed no indication of being due to two compounds. Plots such as that of Figure 3.4 may not be helpful when dealing with mixtures containing many types of functional group, since the number of plates may vary with the chemical type of the solute.

It is more difficult to estimate $\Delta V_{min}/V$ for method (b), because it is much more subject to the effect of variations in chromatographic conditions than is method (a). A measured retention volume depends, among other things, on chart speed, recorder trace thickness, peak shape, injection reproducibility, stability of chromatographic conditions, and column efficiency. An analyst familiar with his particular instrument may be able to decide whether or not two very similar retention times (or volumes) are significantly different. Although such decisions are often to some extent subjective, they can be rationalised, if need be, by repeated injections to establish a mean and a standard deviation (i.e. a distribution curve) for the measured retention volumes of standard and unknown.

Suppose that measurements of the retention volume of the unknown, V_X, and of the standard, V_S, have standard deviations σ_X and σ_S (not to be confused with σ, the standard deviation of a gaussian GC peak). If we make y separate measurements of V_X and V_S, involving $2y$ injections, then the standard error of the difference in the measured average values of the retention volume is given by

$$\sigma_{(\bar{V}_X - \bar{V}_S)} = \frac{\sqrt{(\sigma_X{}^2 + \sigma_S{}^2)}}{\sqrt{y}} \approx \frac{\sigma_S\sqrt{2}}{\sqrt{y}}$$

If $(\bar{V}_X - \bar{V}_S)$ is greater than about three times this standard error, then we may say that the retention volumes of standard and unknown are significantly different. Thus, on this basis,

$$\Delta V_{min} \approx \frac{3\sigma_S\sqrt{2}}{\sqrt{y}}$$

The maximum number of runs which is in practice feasible is about ten for each retention volume (and this is very tedious!), so that if $\sigma_S = 0.01V_S$ (i.e. if retention volumes have a 1% standard deviation, which is typical of a routine instrument*), then $\Delta V_{min}/V \approx 1.3\%$. For a 2% standard deviation in retention volume the corresponding value of $\Delta V_{min}/V$ is 2.7% if ten values of V_S and V_X are measured, while if only five values are obtained this figure rises to 3.8%. It seems therefore that method (b) is comparable with method (a) for packed columns since retention volumes differing by about 1% can be distinguished, provided that a large number of injections are carried out. For capillary columns, method (a) appears to be preferable, except that it may then be difficult to measure the widths of the very narrow peaks to 10%. Method (a) offers the great advantage that only one injection is required. It should be noted that the above treatment is not valid if there is any trend in the measured values of V_X and V_S, due for example to column bleed.

Klein and Tyler[7] have extended the statistical approach to identification from retention data, by taking account of the number of closely eluted substances with which an unknown peak might be confused. If retention values are randomly distributed, the probability of m substances having retention volumes in an interval of given size will be given by a Poisson distribution law:

$$P_m = \frac{\exp(-\rho)\,\rho^m}{m!}$$

* Recent work using computer-controlled injection and data acquisition has shown that over a 30-minute period retention times of around 30 seconds can be measured with a standard deviation of 0.02%.[6]

where ρ is the average occupancy of a large number of intervals. The probability of no substance having a retention volume in a given interval is $\exp(-\rho)$, while the probability of only one substance having a retention volume in the interval is $\rho \exp(-\rho)$. If a peaks can be eluted between retention volumes V_1 and V_2, and we choose an interval of length $\Delta V_{\min}$, as discussed above, then ρ is given by

$$\rho = \frac{a \Delta V_{\min}}{V_2 - V_1} \tag{3.2}$$

It should be noted that we are not concerned with the actual number of peaks from an injected sample, but with the *total possible* number of compounds eluted in a given retention interval. For example, if the sample being analysed is known to contain the elements C, H, and N, the total possible number of eluates, a, would be all those containing these elements. Since $\Delta V_{\min}$ is the smallest difference in retention volume which can be detected, then the probability of simultaneous elution, p_{se}, is simply the probability of two or more substances having retention volumes within the same $\Delta V_{\min}$ interval, i.e.

$$p_{se} = 1 - \exp(-\rho) - \rho \exp(-\rho)$$

Figure 3.5 shows calculated values of p_{se} for various values of ρ. So that the probability of simultaneous elution should be less than, say, 0.05, ρ

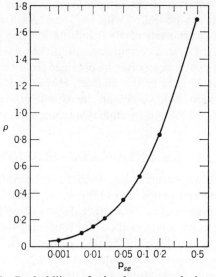

FIGURE 3.5 Probability of simultaneous elution, p_{se}, as a function of the density, ρ, of eluates. (From ref. 7; see text)

must be less than 0·36. If $V_1 = 0$ this means that a must be less than $0·36V_2/\Delta V_{min}$. Now, $100\Delta V_{min}/V_2$ is the percentage accuracy with which the retention volume V_2 can be measured, and so a lower limit for it is about 1% (see above). At the very best, therefore, there have to be fewer than thirty-six possible compounds eluted with retention times less than V_2 if the finding of indistinguishable retention times in the range 0 to V_2 for an unknown and a standard can be taken to imply a 20 : 1 probability that the substances are identical.

In order to reduce mistaken identifications to the one-in-a-thousand level ($p_{se} < 0·001$), a similar argument shows that no more than four possible eluates are allowable. This means that in most practical situations identification at this level of certainty is impossible, given a random distribution of retention volumes. An obvious way to negotiate this difficulty is to reduce ρ and hence p_{se}. From equation (3.2) it can be seen that this requires that either ΔV_{min} or a be reduced. The former involves improving the accuracy of routine retention measurements beyond the 1% level, which is very difficult. On the other hand, a (the number of possible eluates) can be reduced quite easily by making use of additional information about the sample. As an extreme example, if a mixture known definitely to contain only n-alkanes were chromatographed, the chances of mistaken identification under normal conditions would be zero. Useful background information may be provided by almost any analytical technique, elemental analysis of the sample being perhaps the most general. Techniques which can be linked with GC, and which are discussed elsewhere in this book, are particularly helpful in limiting possible identities of unknown peaks. Given a certain amount of such background information, help in identification can often be obtained from retention correlation techniques, discussed later in this chapter. However, if reliance is placed merely on retention behaviour, an increased probability of correct identification can be achieved by employing more than one GC column.

3.2.   RETENTION CORRELATION

Although it is not always practicable, much the best method of identification from retention is to run standards and the unknown under identical conditions on the same apparatus. If a particular standard is not available it may be necessary to rely upon methods for predicting the retention volume of the standard from those of related compounds, and hence upon methods for correlating retention data.

The arrangement, correlation, and extension of retention data are considerably facilitated if use is made of certain relationships which exist between the retention volumes of compounds and some of their physical properties, in particular the vapour pressure, boiling point, and carbon number. The degree of correlation obtained is not always very high, but it is often sufficient to enable the retention volume of an eluate to be estimated, albeit approximately, from known or published retention data without additional experimentation; conversely, it may enable an eluate to be tentatively identified from its retention volume. Although the latter process is in many ways unsatisfactory, it might give *some idea* of an eluate's identity, thereby enabling it to be more exactly pinpointed by some other technique.

Several correlates of the retention volume can be obtained from the retention equation (2.8), i.e.

$$V_g = 273R/\gamma^0 PM \tag{3.3}$$

It should be noted that all methods of correlation apply equally to other retention parameters such as r_{12}, V_N, K, and ΔG^0_{soln} (see Table 2.1).

First, it is evident that V_g and P^{-1}, the reciprocal of the saturated vapour pressure of the solute, should be linearly related for constant γ^0; this is illustrated in Figure 3.6, where the data have been plotted logarithmically for convenience.[8] The slope of the log–log plot predicted by equation (3.3), assuming a constant value of γ^0 for all solutes, is unity, while Figure 3.6 shows that this is so in only two cases (A and D). Furthermore, no known theory enables γ^0 to be predicted for a given system. From a practical point of view, therefore, the correlation of V_g and P^{-1} is only likely to be useful for those systems which are ideal ($\gamma^0 = 1$). Since there is no certain way of knowing in advance whether this is so, this method of correlation is of very limited utility. The systems for which γ^0 is most likely to be unity are those which involve the solution of 'like in like', e.g. alcohols in polyethylene glycol, hydrocarbons in squalane, esters in polyesters, etc. For these systems, log retention against $\log P$ plots can sometimes be useful if the appropriate vapour pressure data are available.

The correlation of V_g with P^{-1} may be removed one step further by considering the relation between P and T_b, the boiling point of the eluate. Thus, from equation (2.9) we have

$$\log_{10} V_g = \log_{10}(273R/M) + 0 \cdot 22kT_b/T - 0 \cdot 22k - 2 \cdot 9 - \log_{10}\gamma^0$$

where k is the Trouton 'constant' (i.e. $\Delta K_{vapn}/T_b$ cal mole^{-1} $^\circ$K^{-1}). It should be noted that k is totally dependent on the solute while γ^0 depends on the solute–solvent system. For a given column and set of conditions, R, M, and T are constant. Hence, for a series of compounds, provided either that $\log \gamma^0$ and k are constant, or that k is constant and $\log \gamma^0$ varies linearly with T_b, a plot of $\log V_g$ against T_b for the series will be a straight line. On the other hand, if either k or $\log_{10} \gamma^0$ *varies* within a series of compounds, and, in the case of the latter, varies non-linearly with T_b, then

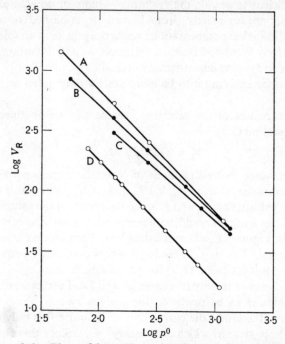

FIGURE 3.6.   Plots of $\log_{10} V_R$ against $\log_{10} p^0$ for: (A) fluoro-, chloro-, bromo-, and iodo-benzenes eluted from benzylbiphenyl at 100 °C; (B) n-butyl-, n-propyl-, ethyl-, and methyl-benzenes, and benzene eluted from 7,8-benzoquinoline at 100 °C; (C) solutes as in B eluted from di-n-propyl tetrabromophthalate at 100 °C; (D) ethyl, n-propyl, and n-butyl alcohols eluted from medicinal paraffin each at several temperatures in the range 30—110 °C. (From ref. 8)

the $\log V_g$ against T_b plot will be non-linear. Furthermore, if different series have the same constant k but different constant γ^0, the $\log V_g$ against T_b plots for the series will be parallel with different intercepts.

Despite a common impression to the contrary, plots of log retention against boiling point for homologous series are almost invariably curved over a large range of T_b, as shown in Figure 3.7 for n-alkanes on a squalane column.[9] Since k for this series is known to be substantially constant at around 20·9, the curvature must be attributed to a variation of $\log \gamma^0$, non-linearly, with T_b. Clearly, if γ^0 is invariant, as with ideal solutions,

$\log V_g$ against T_b plots will consist of a *single* line for all solutes; this is indeed found for eluates which are chemically very similar to the solvent. This situation has also been discussed above in connection with the correlation of V_g and P^{-1}.

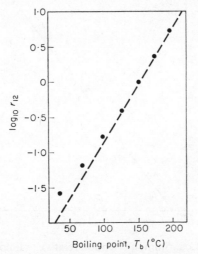

FIGURE 3.7 Logarithm of retention relative to n-nonane plotted against boiling point for n-alkanes from C_5 to C_{11} on squalane at 65 °C. The plot is slightly non-linear (cf. Figure 3.11). (From ref. 9)

The biggest disadvantage of $\log$ retention/T_b correlation is perhaps not that the plots are curved, or that each series falls on a separate line, but that these features are, in addition, unpredictable owing to the lack of a suitable theory for γ^0. In practice this means that if, say, a correlation curve is drawn for ketones, and the unknown eluate is an aromatic material, reliable identification from this curve would probably be impossible. The only case for which this type of correlation can be at all useful is when the identity of the eluate is fairly well known. Figure 3.8 shows the good correlation of relative retention with boiling point for twenty terpenes on Apiezon-L, while Figure 3.9 shows the $\log r_{12}$ against T_b plot for a wide variety of hydrocarbons on a non-selective solvent. It can be seen that, within reasonable limits of error, an eluate known to be a hydrocarbon can be assigned a boiling point, and hence a provisional identity from its retention volume. However, the method would be ineffectual if the eluate were less well defined.

Boiling point relationships are sometimes useful when using linear temperature programmed GC. For example,[12] the retention times (or

temperatures) of a variety of alkylated aromatic hydrocarbons on a
ditridecyl phthalate column correlate linearly with their boiling points, as
shown in Figure 3.10. A separate line is obtained for all the isomers of a

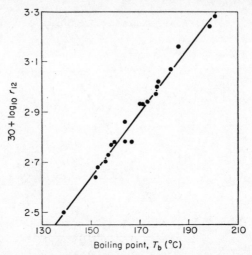

FIGURE 3.8   Logarithm of retention relative to limonene plotted
against boiling point for various acyclic, monocyclic, and bicyclic
terpenes on Apiezon-L at 120 °C. (From ref. 10)

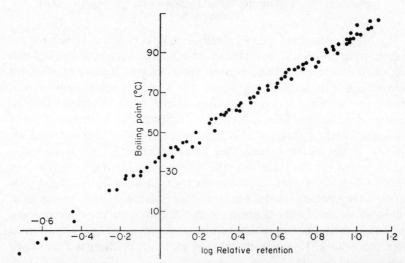

FIGURE 3.9   Correlation of the logarithm of relative retention with boiling
point for a variety of alkanes, alkenes, alkadienes, alkynes, cycloalkenes, and
cycloalkadienes on tetraamylsilane (3% w/w) at 25 °C.   (From ref. 11)

given carbon number (demonstrated for $z = 8$ to 10). Since these lines overlap, a given retention time does not correspond to a unique boiling point. Nonetheless, such correlations are useful aids to the interpretation

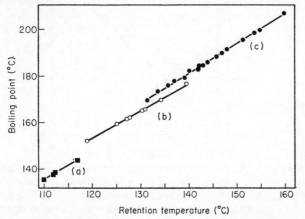

FIGURE 3.10   Plot of retention temperature against boiling point for various alkyl-substituted benzenes eluted from a linear programmed temperature ditridecyl phthalate capillary column. (a) C_8H_{10} isomers; (b) C_9H_{12} isomers; (c) $C_{10}H_{14}$ isomers. (From ref. 12)

of the results of supplementary techniques such as mass spectrometry, particularly when dealing with a large number of closely related compounds.

Another method of correlation, which is applicable only to homologous series, is the relation between log retention and z, the number of atoms (usually carbon) in the series chain. Figure 3.11 shows an example for n-alkanes. It is clear that the linear correlation is very much better than that of Figure 3.7 for $\log r_{12}$ against T_b which is based on the same retention data. In general, plots of log retention against z exhibit a high degree of linearity for all but the first few members of the series, and have been obtained for many different types of compound. Figure 3.12 shows such plots for adjusted retentions (Table 2.1) of n-alkanes, n-alcohols, and acetates on a porous polymer packing. Since, from equation (2.7):

$$\ln V_g - \ln (273R/M) = -\Delta G^0_{soln}/RT$$

it follows from the above that, for a given series, $\Delta G^0_{soln} = bzRT + a$, where a and b are constants. It is found experimentally that related series frequently have parallel $\log V_g$ against z plots, and hence the same value of b, but various values for a. This result may be interpreted in terms of the free energy of solution of a CH_2 group, $\Delta G^0_{soln}(CH_2)$, as follows:

$$\Delta G^0_{soln}(C_z - X) = z\Delta G^0_{soln}(CH_2) + a$$

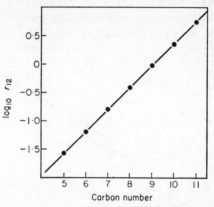

FIGURE 3.11   Logarithm of retention relative to n-nonane plotted against carbon number, for n-alkanes from C_5 to C_{11} on squalane at 65 °C. The retention data are the same as for the boiling-point plot of Figure 3.7, but this carbon-number plot is a better straight line. (From ref. 9)

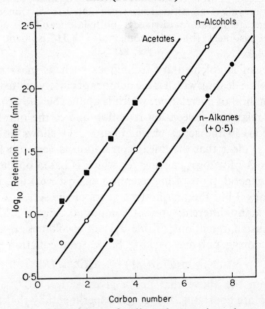

FIGURE 3.12   Logarithm of adjusted retention times plotted against carbon number for n-alkanes, n-alcohols, and n-alkyl acetates on a porous copolymer of styrene and divinylbenzene (Chromosorb 102) at 200 °C. (From ref. 13)

The constant a may be broken down into $\Delta G^0_{soln}(X)$, the free energy associated either with the functional group or with some structural feature of the particular series, and a', another constant, as follows:

$$a = \Delta G^0_{soln}(X) + a'$$

Hence

$$\Delta G^0_{soln}(C_z - X) = z\Delta G^0_{soln}(CH_2) + \Delta G^0_{soln}(X) + a'$$

As shown in Figure 3.13, correlation diagrams for phenylalkanes[14] comprise several parallel lines, each of which refers to a particular position

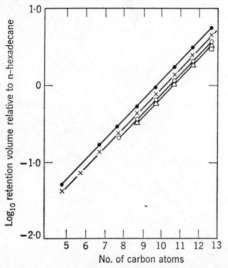

FIGURE 3.13   Relationship between log retention volume relative to n-hexadecane and the number of carbon atoms in the chain of various phenyl alkanes eluted from Apiezon-L at 130 °C.
●, 2-Phenyl isomers; ×, 3-phenyl isomers; ○, 4-phenyl isomers; △, 5- and 6-phenyl isomers; □, 7-phenyl isomers.
(From ref. 14)

of substitution by the phenyl group. In this instance, the value of $\Delta G^0_{soln}(X)$ for each line is associated with the form of the carbon skeleton of the corresponding eluates. Similar results are obtained for many other systems.

A corollary of the parallelism of the lines in Figure 3.13 is that the logarithms of the retention volumes of phenylalkanes containing the same number of carbon atoms change by a fixed increment (the retention volumes themselves by a fixed *factor*) as the position of substitution varies. This sort of relation has been found to hold for numerous compounds and their variants (see, e.g., n- and iso-fatty acids and their methyl esters,[15] and dimethyl acetals of fatty acid aldehydes[16]), and is the basic observation which has led to the formulation of structural rules for the calculation of retention volumes, particularly in connection with retention indices.

Correlation of log retention with z applies also to atoms other than carbon, for example, sulphur.[17] An interesting example is shown in Figure 3.14, which illustrates the retention behaviour of straight-chain

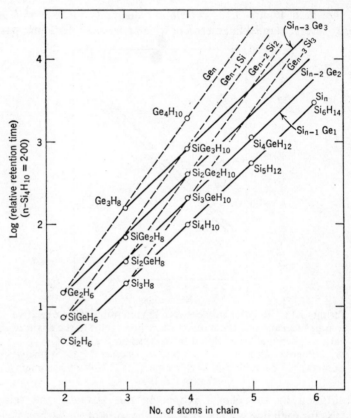

FIGURE 3.14   Retention correlation grid for silanes, germanes, and silicogermanes eluted on Silicon 702 at 19 °C. (From ref. 18)

hydrides of silicon and germanium.[18] Many of these compounds are not available as standard samples, and are difficult to prepare in the pure state. Also, they are spontaneously inflammable in air, and hence GC is an ideal technique for studying them. The difficult problem of identification was solved by a variety of techniques, including measurement of the molecular weights of the peaks by means of a gas-density balance and katharometer (cf. Chapter Seven). Reference to Figure 3.14 shows that the resultant correlation diagram consists of a grid such that for a fixed number of silicon atoms the logarithm of the retention volume increases by 1·07 units for the addition of each GeH_2 unit. Similarly, for a fixed number of germanium atoms, the logarithm of the retention volume increases by 0·72 log unit for the addition of each SiH_2 unit. This pattern enables prediction of the retention volumes of many compounds other than those which were actually measured.

Finally, mention should be made of the correlation of retention data between two columns. This is generally done by plotting logarithmically the retention data against each other. Then, provided that one of the columns is selective, various classes of compound will lie in characteristic regions of the diagram. Figure 3.15 shows a logarithmic cross-plot of the

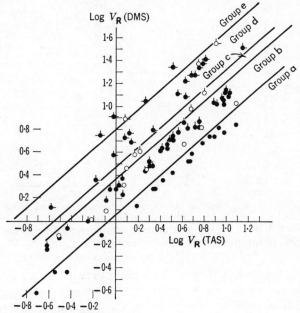

FIGURE 3.15   A logarithmic cross-plot of the retention data of a large number of hydrocarbons on tetraamylsilane (TAS) and dimethylsulpholan (DMS) columns at 25 °C.

●, Alkanes; ◖, alkenes; ◗, alkadienes;─◖, alkynes; ○, cyclo-alkanes; ◔, cycloalkenes; ◓, cycloalkadienes. (From ref. 11)

retention volumes of a large number of hydrocarbons on tetraamylsilane and dimethylsulpholan columns.[11] Distinct regions which enclose all the members of certain classes can be discerned: (a) alkanes; (b) alkenes and

C

cycloalkanes; (c) cycloalkenes; (d) alkadienes and cycloalkadienes; and (e) alkynes. In effect, moving across the diagram from bottom right to top left involves crossing areas of increasing unsaturation, owing to the higher retention of such materials on dimethylsulpholan. Unambiguous definition of the class of hydrocarbon in (b) and (d) is possible on the basis of post-column hydrogenation experiments.[19] If, for example, an unidentified eluate is known to be in group (b), then hydrogenation will change its retention volume only if it is an alkene.

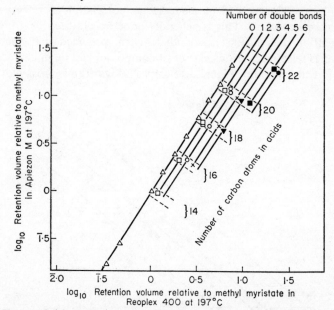

FIGURE 3.16   Log–log plot of retention data relative to methyl myristate for unsaturated acids at 197 °C on Apiezon-M and Reoplex 400 (polypropylene glycol adipate). (From ref. 20)

Similarly, Figure 3.16 shows a cross-plot of retention data for a variety of unsaturated acids on Apiezon and polypropylene glycol adipate columns.[20] The position of a point on the diagram not only gives the number of double bonds but also gives the number of carbon atoms in the acid. Cross-plotting can be extended to three columns by using triangular graphs.[21] Clearly, such methods are very powerful when identifying closely related materials.

Another method of treating the retention data from more than one column will be given below in the section on Kovats indices.

Similar methods of correlation can also be applied to other techniques such as coupled GC and thin-layer chromatography.[22]

3.3. THE PRESENTATION OF RETENTION DATA

So far in this chapter discussion has centred round experimental or specific retention volumes. There are, however, several disadvantages in dealing with such measurements. Experimental retention volumes depend, of course, upon various column parameters such as flow rate, temperature, and weight of liquid phase, and cannot in general be directly compared with data from the literature. Specific retention volumes cannot be calculated without knowing accurately the weight of liquid phase on the column, and even with the greatest care errors cannot readily be reduced below 1%. These and other problems are generally overcome by scaling all experimental retention data according to the retention volume of at least one eluate which is included in the literature data. If several eluates are used for this purpose it is convenient to cross-plot the experimental and published data, when a straight line should be obtained with a slope which is equal to the required scaling factor. In effect, if only one reference eluate is used, retention data are converted to *relative* retention data, and ideally the effect is that differences in column conditions between the two sets of data are automatically allowed for. In practice the procedure is far from accurate. It should be noted that even in principle this technique is justified only if experimental and published data refer to the same temperature, and also if any solid support retention is identical.

Another disadvantage of simple retention volumes is that they have a strong temperature dependence, reflecting the heat of solution of the eluate. Relative retention data tend to be less temperature-sensitive since they depend on the *difference* in the heats of solution of the two eluates.

$$\log r_{12} = \log (V_{N(1)}/V_{N(2)})$$
$$= -(\Delta G^0_{soln(1)} - \Delta G^0_{soln(2)})/RT$$
$$= -\Delta(\Delta G^0_{soln})/RT$$
$$= -\Delta(\Delta H^0_{soln})/RT + \Delta(\Delta S^0_{soln})/R$$

However, it is evident from these equations that, unless solutes 1 and 2 have almost equal heats of solution $[\Delta(\Delta H^0_{soln}) = 0]$, r_{12} will vary significantly with T. Indeed, it is often observed that the order of elution of solutes is reversed by a change of temperature.[23]

McCrea and Purnell[24] have shown that temperature-independent retention can be obtained for certain systems in which the volume of the stationary phase, V_L, increases with temperature at a reasonable rate. Since partition coefficients invariably *decrease* as the temperature is raised, conditions can be found in which the retention volume of a solute given by

$$V_N = KV_L$$

becomes temperature-independent. Using a column of methyl stearate in n-octadecanol on Chromosorb-P, these authors obtained temperature-independent retention of a variety of alkanes between 44·9 and 54·1 °C, by running the column in series with a conventional squalane column. A phase change in the 'compensating' methyl stearate—octadecanol column occurs over an extended temperature range, causing V_N to *increase* with increasing temperature, in opposition to the normal decreases in V_N on the squalane column.

If such systems can be devised with different selectivities, they will undoubtedly gain much favour in analytical work, since temperature control is one of the major factors determining the reproducibility of retention data.

Some authors have proposed that retention data should be reported relative to one retention volume. One method, the R_{x9} system, involves the assignment to n-nonane of a theoretical retention volume, V_9', on the basis of the line of best fit through data obtained for n-alkanes plotted as $\log V_N$ against z.[9] The R_{x9} value for any eluate is given by

$$R_{x9} = V_x/V_9'$$

where V_x is its retention volume. The main disadvantages of the R_{x9} values are that, in common with most relative retention volumes, they are often significantly temperature-dependent and they cannot be accurately obtained when V_x and V_9' are widely different. Furthermore, laboratory trials[25] have shown that they are less reproducible than other indices of retention, which are discussed below.

Ideally, what is required for the presentation of relative retention data is a number of standard solutes covering a wide range of retention, the retention volumes for which correlate in some convenient way. Then, any column can be standardised from a single injection of a mixture of standards, and the retention volume of any particular eluate can be compared with its *nearest*-eluted standard. This is precisely how the Kovats index system[26] is formulated, using n-alkanes as its standards, and correlating retention volumes with z. The system is widely used, and has been endorsed by various authorities.[25]

The retention index of n-alkanes under all conditions is arbitrarily fixed at $100z$, and the retention index of any other material is obtained by assigning to it a hypothetical carbon number on the basis of its retention relative to the nearest-eluted n-alkane, and multiplying it by 100. Hence, if a material with net retention volume V_{Nx} is eluted between alkanes containing z and $z+1$ carbon atoms,

$$I_x = 100\left[z+\left(\frac{\log V_{Nx}-\log V_{Nz}}{\log V_{N(z+1)}-\log V_{Nz}}\right)\right] \tag{3.4}$$

Figure 3.17 illustrates graphically the derivation of I_x. It should be noted that, provided that there is a linear variation of $\log V_{Nz}$ with z, I_x is simply related to r_{xz}, the relative retention of X and C_zH_{2z+2}, by the equation

$$I_x = 100(z+\log r_{xz}/b)$$

where b is the slope of the $\log V_{Nz}$ against z plot. In particular, R_{x9} and I_x values are related by the equation

$$I_x = 100(9+\log R_{x9}/b)$$

It can be seen from these equations that values of r_{xz} can only be obtained from I_x if b is known. In addition, since $r_{xz} = V_{Nx}/V_{Nz} = K_x/K_z$, where K_x and K_z are the relevant partition coefficients, it is evident that K_x, the quantity of thermodynamic significance (at the column temperature), can be computed only if K_z is known, and hence if V_{Nz} ($=K_zV_L$) and V_L are known. It follows that, if the extensive data now reported as Kovats retention indices are to be of maximum benefit, they should be given together with the following data:

(i) b, the slope of the $\log V_{Nz}$ against z plot;

(ii) V_{Nz}, the net retention volume of the n-alkane with z carbon atoms;

(iii) the volume or weight of liquid phase;

(iv) the column temperature.

In addition, for analytical purposes, all other column conditions and full details of the stationary phase should be given. It is also helpful if some idea of time scale is indicated by giving the retention *time* of one material.

The value of b for any column depends only on the nature of the retention process. This is easily shown as follows. Since $V_{Nz} = K_zV_L$,

$$b = \partial/\partial z(\log V_{Nz}) = \partial/\partial z(\log K_z) = -[\partial/\partial z(\Delta G^0_{\text{soln}(z)})]/2\cdot303RT$$

[see equation (2.7)]. Hence, at a given temperature, b is dependent only on the variation of $\Delta G^0_{\text{soln}(z)}$, the standard free energy of solution of C_zH_{2z+2}, with z. This is a fundamental thermodynamic quantity completely independent of anything except the nature of the retention process. Hence, if published retention indices for a given column are to be used, it is first of all necessary to make certain that the equivalent column which will have been prepared gives the same value of b as the given column. This comparison will, of course, have to be made at the same temperature. It can be shown theoretically, and is confirmed by experiment, that I_x varies more slowly with temperature than do relative retention data. Over small ranges of temperature, I_x is often assumed to vary linearly with T or $1/T$, although the evidence is not very strong[27] and theory predicts a more complicated dependence. Values of $\partial I_x/\partial T$ calculated on the assumption of a linear variation with T are typically around 0.05% per $C°$, although much higher and lower values have been obtained. In order to see more clearly how I_x is affected by temperature it is necessary to rewrite equation (3.4) in thermodynamic terms, as follows:

$$\frac{I_x}{100} = z + \left(\frac{\Delta G^0{}_x - \Delta G^0{}_z}{\Delta G^0{}_{z+1} - \Delta G^0{}_z}\right)$$

where ΔG^0 represents the standard free energy of solution. Now, for all but the smallest n-alkanes $\Delta G^0{}_z/\Delta G^0{}_{z+1} = z/(z+1)$. Hence,

$$\frac{I_x}{100} = z + \left(\frac{\Delta G^0{}_x - \Delta G^0{}_z}{\Delta G^0{}_z/z}\right) = \frac{z\Delta G^0{}_x}{\Delta G^0{}_z} = \frac{z(\Delta H^0{}_x - \Delta S^0{}_x T)}{\Delta H^0{}_z - \Delta S^0{}_z T}$$

It can be seen from this equation that the variation of I_x with T is complicated and cannot be stated without reference to the particular. In any event, the assumption of a simple linear variation, whilst convenient, is unlikely to be accurate, or indeed significant.

3.4 IDENTIFICATION BY MEANS OF THE KOVATS INDEX

For purposes of identification, retention indices are handled in much the same way as retention volumes.[28] Retention coincidence, in terms of I-values, between standard and unknown eluates is sought; structural relations of I are used to elucidate a peak's identity; differences in the retention index of an eluate on two different types of column are compared with published data; and the variation of I with temperature is considered.

One of the most important advantages of the Kovats index is that it is more reproducible than other forms of retention data.[25,29] Hence, provided that the column used, and its temperature, are the same as for the reported data, measured values of I can be compared with literature values with some confidence. In these circumstances, retention coincidence without the need for the appropriate standard materials can be effected more reliably. However, if the standard itself is used, it is pointless to convert retention data to Kovats indices since these data can be compared directly.

The procedure is, first, to measure the retention volumes of a series of n-alkanes injected as a mixture, and chosen to extend throughout the region in which the unknown solutes are eluted. The latter are then chromatographed and their Kovats indices calculated from their retention volumes as indicated in Figure 3.17. The values obtained can then be compared with the ever-growing body of literature reporting retention data as Kovats indices.

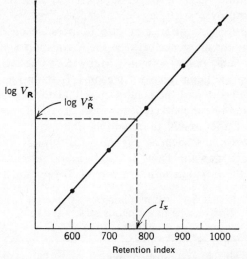

FIGURE 3.17   Diagrammatic illustration of the derivation of the Kovats index, I_x, of a material eluted between n-heptane and n-octane

Studies of the variation of the Kovats index with structure[30,31] have shown that there is an approximately constant difference between the values for a parent solute and its derivative. Swoboda[32] has called this difference the 'functional retention index', and some values due to him are shown in Table 3.1. The use of this index in connection with carbon skeleton chromatography (p. 103) is particularly attractive.

TABLE 3.1   Some values of functional retention indices[32]
(parent solute: n-alkane)

Class of compound	Squalane (75 °C)	Dinonyl phthalate (74 °C)
n-Alkanol	230	402
n-Alkanal	152	288
n-Alk-2-enal	203	378
n-Alka-2,4-dienal	248	450

Many other structural correlations have been observed (e.g. refs. 33, 34), and it is not inconceivable that a set of rules can be found which would enable the Kovats index of any compound to be calculated from its structure. However, it is likely that to achieve any reasonable accuracy the number of structural parameters needed would be so large as to render the scheme inoperable.[35] Certain simple rules, such as those mentioned above, can nonetheless provide useful guides to identity.

A preliminary classification may also be possible on the basis of ΔI values, ΔI being the difference between the Kovats index of a given solute on 'non-polar' and 'polar' columns. Many ΔI values have been determined[36,37] using the liquid phases Apiezon-L (paraffin) and Emulphor-O (polyethylene glycol), and the results show that the members of a given class of compound have characteristic values of ΔI. Such relationships are restatements of the trends in retention data which are apparent in logarithmic cross-plots such as Figure 3.15. Typical values of ΔI for various series are shown in Table 3.2, and general rules have been devised to account for the variation with structure of ΔI values in a given series.[37]

TABLE 3.2 I(Emulphor-O) $- I$(Apiezon-L)
for various homologous series[a] at 130 °C
(ref. 37)

Nitriles and nitro-compounds	340—410
Alcohols	300—360
Formates	200—280
Acetates	160—200
Butyrates	180—250
Aldehydes and ketones	170—260
Bromides and chlorides	120—180
Ethers	60—100
Olefins	30—50
Naphthenes	20—50

[a] Aromatic compounds typically have values 160–200 higher.

TABLE 3.3   Variation of I with temperature on
polyglycol columns[38]

Class of compound	$- \partial I/\partial(1/T) \times 10^{-2}$
Alkanes[a]	0 ± 4
Alk-1-enes[a]	14 ± 3
Alkenes except alk-1-enes[a]	15—24
Monocyclic compounds[b]	70—120
Bicyclic compounds[b]	170—220

[a] On Carbowax 20M. [b] On Emulphor-O.

The rate of variation of retention data with temperature may be of use in the identification of an eluate. Indeed, differences between eluates are often sufficient to reverse orders of elution over quite small ranges of temperature. For example, on a squalane column at 46 °C 2,2-dimethyl-3-ethylpentane is eluted before 2,6-dimethylheptane, whereas at 106 °C the order is reversed. Similarly, on the same substrate, cycloheptane precedes n-octane at 58 °C and follows it at 70 °C.[23] Table 3.3 shows some values of $\partial I/\partial(1/T)$ obtained for a variety of hydrocarbons on polyglycol columns.[38] Whilst it is not possible to base identification solely on data such as these, it would be perfectly reasonable to use them in confirmation of an identity established by other means. The high temperature-dependence of the retention indices of cyclic materials is particularly noteworthy.

If the Kovats index becomes the standard way of reporting retention data, a great deal will have been achieved. However, it should never be forgotten that the index does not provide anything which is not available from straightforward retention volumes. At present, its use suffers from the real disadvantage that there is a tendency to attach too much importance to the indices themselves at the expense of obscuring the theory and practice from which they are derived. Whether retention indices are used or not, comparison of the retention of unknown solutes with standard samples on several columns with substantially different types of liquid phase should be used wherever possible in preference to inexact rules applied to a single column.

REFERENCES

1. *Compilation of Gas Chromatographic Data* (ed. J. S. Lewis), ASTM, Philadelphia (1963).
2. *Gas Chromatographic Retention Data Compilation* (ed. O. E. Schupp and J. S. Lewis), 2nd Edn., ASTM, Philadelphia (1967).
3. W. O. McReynolds, *Gas Chromatographic Retention Data*, Preston Technical Abstracts Co., Evanston, Ill., U.S.A. (1966).
4. V. Cejka, M. H. Dipert, S. A. Tyler, and P. D. Klein, 'Detection and analysis of unresolved multiplet chromatographic peaks', *Analyt. Chem.*, **40**, 1614 (1968).
5. P. John, private communication.
6. J. E. Oberholtzer and L. B. Rogers, 'Precise gas chromatographic measurements', *Analyt. Chem.*, **41**, 1234 (1969).
7. P. D. Klein and S. A. Tyler, 'Accuracy in determination of chromatographic mobility and its significance in identification of compounds', *Analyt. Chem.*, **37**, 1280 (1965).
8. H. Purnell, *Gas Chromatography*, Wiley, New York (1962), p. 209.

9. M. B. Evans and J. F. Smith, 'Gas–liquid chromatography in qualitative analysis. II. The reproducibility of retention data in R_{x9} units', *J. Chromatog.*, **6**, 293 (1961).

10. M. H. Klouwen and R. Ter Heide, 'A systematic analysis of monoterpene hydrocarbons by gas–liquid chromatography', *J. Chromatog.*, **7**, 297 (1962).

11. B. Smith, R. Ohlson, and G. Larson, 'Relative retention times of saturated and unsaturated hydrocarbons and their use for the determination of boiling point and hydrocarbon type', *Acta Chem. Scand.*, **17**, 436 (1963).

12. D. E. Willis, 'Retention time-boiling point correlations during programmed temperature capillary column analysis of C_8-C_{12} aromatic compounds', *Analyt. Chem.*, **39**, 1324 (1967).

13. Perkin-Elmer Analytical News, No. 2 (1968).

14. J. M. Blakeway and D. B. Thomas, 'Gas–liquid chromatography of linear detergent alkylates', *J. Chromatog.*, **6**, 74 (1961).

15. H. Brandenberger and S. Müller, 'A gas chromatographic separation of the volatile fatty acids of black tea', *J. Chromatog.*, **7**, 137 (1962).

16. G. M. Gray, 'The separation of the long chain fatty aldehydes by gas–liquid chromatography', *J. Chromatog.*, **4**, 52 (1960).

17. D. M. Oaks, H. Hartmann, and K. P. Dimick, 'Analysis of sulfur compounds with electron capture/hydrogen flame dual channel gas chromatography', *Analyt. Chem.*, **36**, 1560 (1964).

18. C. S. G. Phillips and P. L. Timms, 'Some applications of gas chromatography to inorganic chemistry', *Analyt. Chem.*, **35**, 505 (1963).

19. B. Smith and R. Ohlson, 'Hydrogenation as an aid in the identification of unsaturated hydrocarbons by gas chromatography', *Acta Chem. Scand.*, **14**, 1317 (1960).

20. A. T. James, 'Determination of the degree of unsaturation of long chain fatty acids by gas–liquid chromatography', *J. Chromatog.*, **2**, 552 (1959).

21. I. Brown, 'Identification of organic compounds by gas chromatography', *Nature*, **188**, 1021 (1960).

22. J. Janák, 'Multi-dimensional chromatography using different developing methods', *J. Chromatog.*, **15**, 15 (1964).

23. M. C. Simons, D. B. Richardson, and I. Dvoretzky, 'Structural analysis of hydrocarbons by capillary gas chromatography in conjunction with the methylene insertion reaction', in *Gas Chromatography 1960* (ed. R. P. W. Scott), Butterworths, London (1960), p. 211.

24. P. F. McCrea and J. H. Purnell, 'Temperature independent retention in gas chromatography', *Nature*, **219**, 261 (1968). See also 'Gas chromatographic column systems exhibiting temperature independence of solute retention', *Analyt. Chem.*, **41**, 1922 (1969).

25. Institute of Petroleum GC Discussion Group Data Sub-committee, 'Recommendations for the publication of retention data', *Gas Chromatography 1964* (ed. A. Goldup), Institute of Petroleum, London (1965), p. 348; *J. Gas Chromatog.*, **3**, 298 (1965); **4**, 1 (1966).

26. E. Kovats, 'Gas chromatographic characterization of organic substances in the retention index system', in *Advances in Chromatography*, Vol. 1 (ed. J. C. Giddings and R. A. Keller), Marcel Dekker, New York (1965), p. 229.

27. R. A. Hively and R. E. Hinton, 'Variation of the retention index with temperature on squalane substrates', *J. Gas Chromatog.*, **6**, 203 (1968).

28. A. Švob, D. J. Deur-Šiftar, and V. Jarm, 'Gas chromatographic separation and identification of dimerization products of α-methylstyrene', *J. Chromatog.*, **38**, 326 (1969).
29. Institute of Petroleum GC Discussion Group Data Sub-committee, 'The fatty acid ester–hydrocarbon correlation trial', *Gas Chromatography 1966* (ed. A. B. Littlewood), Institute of Petroleum, London (1967), p. 395.
30. J. Zulaica and G. Guiochon, 'Separation and identification of aliphatic and aromatic di-esters and phosphate tri-esters by vapour phase chromatography', *Bull. Soc. Chim. France*, 1242 (1963).
31. C. Baron and B. Maume, 'Gas chromatographic behaviour and stereochemical structures of menthol, menthane, borneol, menthoglycol, and their stereoisomers, and camphor', *Bull. Soc. Chim. France*, 1113 (1962).
32. P. A. T. Swoboda, 'Quantitative and qualitative analysis of flavour volatiles from edible fats', in *Gas Chromatography 1962* (ed. M. van Swaay), Butterworths, London (1962), p. 273.
33. C. Merritt, J. T. Walsh, D. H. Robertson, and A. I. McCarthy, 'Qualitative gas chromatographic analysis by means of retention volume constants— behaviour of isomers', *J. Gas Chromatog.*, **2**, 125 (1964).
34. A. I. McCarthy, H. Wyman, and J. K. Palmer, 'Gas chromatographic identification of banana fruit volatiles', *J. Gas Chromatog.*, **2**, 121 (1964).
35. P. G. Dodsworth, contribution to discussion after ref. 32, p. 287.
36. E. Kovats, 'Gas chromatographic characterization of organic compounds. I. Retention indices of aliphatic halides, alcohols, aldehydes and ketones', *Helv. Chim. Acta*, **41**, 1915 (1958).
37. A. Wehrli and E. Kovats, 'Gas chromatographic characterization of organic compounds. III. Calculation of retention indices for aliphatic, alicyclic and aromatic compounds', *Helv. Chim. Acta*, **42**, 2709 (1959).
38. H. Widmer, 'Gas chromatographic identification of hydrocarbons using retention indices', *J. Gas Chromatog.*, **5**, 506 (1967).

CHAPTER FOUR

SELECTIVE ABSTRACTION AS A MEANS OF IDENTIFICATION

A reactor which specifically removes one or more classes of compound is obviously very useful in qualitative analysis. This chapter considers such abstractors, both physical and chemical, and gives details of the various methods of identification which rely upon them. Abstraction usually takes place on-stream, the active reagent being distributed on a convenient solid support material.

Also considered in this chapter are methods of pre-column abstraction in which the eluate is distributed between two immiscible liquids prior to analysis, and is identified by its partition coefficient between them. Other methods of abstraction, in which the rate of reaction of an eluate with some reagent assists in its identification, are considered in Chapter Five in the context of reaction gas chromatography (p. 105). A discussion of the use of the piezoelectric sorption detector in reactive abstraction is given in Chapter Eight (p. 152).

With certain notable exceptions, such as the use of molecular sieves to separate branched and straight-chain hydrocarbons, abstraction GC has not been extensively used for the identification of GC peaks, although it has the important advantages of simplicity and cheapness. Its main disadvantage is that it may involve a great deal of preliminary experimentation, and it is perhaps most useful for the qualitative analysis of a large number of different samples of a similar nature, particularly when complex mixtures are involved. Then, a complicated chromatogram can be considerably simplified, and guidance given to its interpretation by the use of suitable abstractors.

4.1 COMPLETE ABSTRACTION

In conventional qualitative analysis the presence or absence of a particular functional group is usually indicated by a distinctive event such as a colour change, precipitation, or the evolution of a characteristic gas when a specific reagent is added. In qualitative GC analysis, a reagent can be used to remove peaks due to a given class of compound from the chromatogram provided that the products of reaction have low volatility. The nature of this reaction may be either physical or chemical.

For example, frequent use is made of concentrated sulphuric acid to remove olefins from a sample containing saturated compounds, and of molecular sieve to separate straight and branched-chain alkanes. There are many other selective abstractors reported in the literature (see Table 4.1), and undoubtedly many others await discovery, since the demands on the reagent are rather different from those of conventional analysis.

TABLE 4.1 Some abstraction reagents suitable for use in qualitative GC[a]

Abstractor	Materials removed	Ref.
Molecular sieve 5A	n-Alkanes and other straight-chain molecules	1
Mercuric acetate–Hg(NO$_3$)$_2$–ethylene glycol[b]	Alkenes	2
Maleic anhydride–silica gel	Dienes	3
1 : 1 mixture of 1M-Hg(ClO$_4$)$_2$ and 2M-HClO$_4$[b]	Alkenes, alkynes	4
20% HgSO$_4$–20% H$_2$SO$_4$[b]	Alkenes, alkynes	5
4% AgNO$_3$–95% H$_2$SO$_4$[b]	Aromatics, alkenes, alkynes	5
Concentrated H$_2$SO$_4$[b]	Aromatics, alkenes, alkynes	1
Molecular sieve 10X	Aromatics	6
Boric acid[b]	Alcohols	7, 10
NaOH–quartz	Phenols	8
FFAP[b]	Aldehydes	9, 10
NaHSO$_3$–ethylene glycol[b]	Aldehydes	2
Benzidine[b]	Aldehydes, ketones	10
o-Dianisidine[b]	Aldehydes	10
Phosphoric acid[b]	Epoxides	10
Zinc oxide	Acids	10
Molecular sieve 5A	Aldehydes (from ketones), acids (from esters)	11
Versamid 900[b]	Alkyl halides, phenyl halides, and fatty acid α-bromo-esters (labile halogen atoms)	12
NaBr–alumina	Organic compounds with functional groups (ketones, alcohols, etc.)	13
AgNO$_3$–alumina ⎫ CuCl–alumina ⎭	As above, plus alkenes, alkynes, and aromatics	13

[a] Care should be exercised in the use of these reagents, and their characteristics should always be checked with those of standard materials before application.
[b] This symbol indicates that the abstraction reagent is distributed on firebrick or similar support.

4.1a Experimental methods

Any rapid selective physical or chemical abstraction process can be used, but it is obvious that one must have some idea of the sort of components

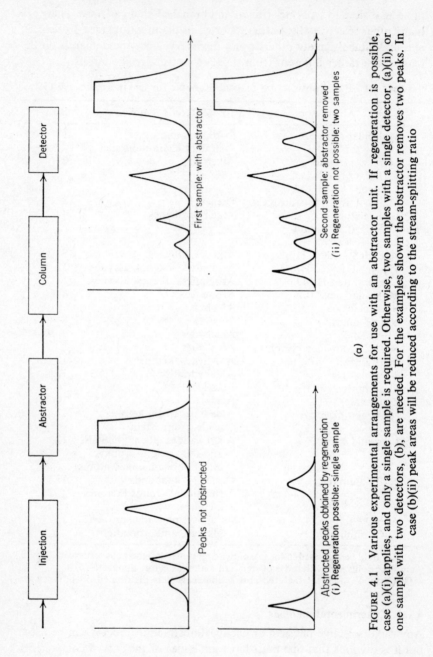

(a)

FIGURE 4.1   Various experimental arrangements for use with an abstractor unit. If regeneration is possible, case (a)(i) applies, and only a single sample is required. Otherwise, two samples with a single detector, (a)(ii), or one sample with two detectors, (b), are needed. For the examples shown the abstractor removes two peaks. In case (b)(ii) peak areas will be reduced according to the stream-splitting ratio

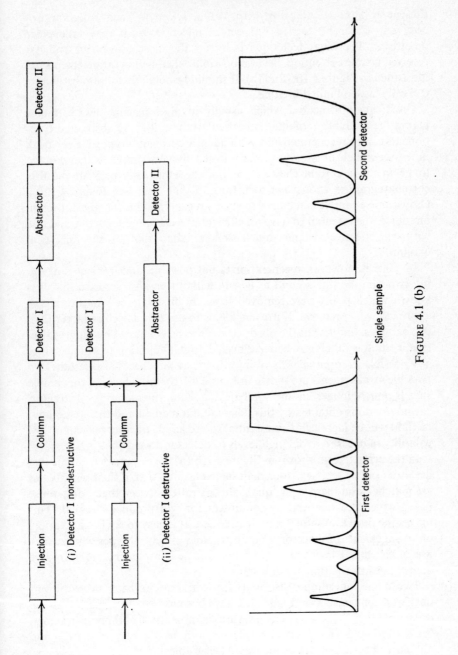

FIGURE 4.1 (b)

present so that a suitable abstractor can be selected. The abstraction process may either be carried out before injection, or it may take place on-stream. The latter technique is by far the most convenient and has received greatest attention. Several possible experimental arrangements are illustrated in Figure 4.1(a) and (b). It should be noted that most abstractors can only have a limited lifetime.

On-stream abstraction, which usually involves placing the abstractor before the analytical column, is particularly useful if the abstracted components can be regenerated and subsequently chromatographed. Each peak in the second chromatogram, which is usually very simple, is then known to belong to a certain class of compound, while each peak in the first chromatogram is known *not* to belong to that class [see Figure 4.1(a)]. Only a single sample is required in such a regeneration technique, although of course it is essential to have an efficient pre-column abstractor, and one with an effective volume much smaller than that of the analytical column.

In general, however, regeneration is not possible, and it is necessary to use two samples, the second to be run without an abstractor so that it is known which peaks were removed from the first sample. An alternative approach, when removal is irreversible, is to use two detectors rather than two samples, and to put the abstractor *after* the column. This single-injection method is inherently better than the double-injection technique, since any possible irreproducibility of injection is avoided; it is also faster. Two possible arrangements are shown in Figure 4.1(b). Provided that the sample spends much longer in the analytical column than in the abstractor, chromatograms obtained by this post-column method and the pre-column method will be identical. It is essential to use an abstracting reagent of low volatility in order to avoid high levels of detector noise.

In the arrangement shown in Figure 4.1(b)(ii) the responses from the two detectors can be set back-to-back in such a fashion that, if their sensitivities are matched and there is an equal splitting ratio between them, only those peaks which are removed by the abstractor will produce a signal. This makes the device a useful yes/no instrument. It is easy to envisage a variety of arrangements of detectors and abstractors of several types which make use of 'chemical logic' in a more complicated fashion in order to give qualitative information about a sample.

Two flame ionisation detectors (FID) with back-to-back electrical outputs, and connected as in Figure 4.2, have been used to distinguish between classes of hydrocarbon.[5] By using various combinations of three abstractors, the components of the detector response due to alkanes, aromatics, and aliphatic unsaturates can be separated (see Table 4.2).

In another, similar approach, a sample is passed through a column, then through a reaction zone (reactor or abstractor) and detector into a cold trap. The abstractor is then changed and the contents of the cold trap are

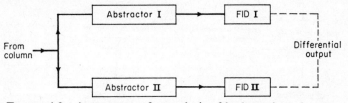

FIGURE 4.2   Arrangement for analysis of hydrocarbons by type (see Table 4.2). The detector outputs are connected back-to-back. (From ref. 5)

re-injected. The process is repeated several times with the same sample, using different abstractors (and reactors), and the results are analysed in terms of a flow diagram as shown in Figure 4.3. Figure 4.4 shows chromatograms obtained by working down the left-hand side of Figure 4.3.

TABLE 4.2   Determination of hydrocarbon types obtained by splitting effluent between two sets of abstractor and detector[5] (see Figure 4.2)

Abstractor I[a]	Abstractor II	Components of sample contributing to detector response[b]		Differential output
		Detector I	Detector II	
SSSA	MSSA	P	P + Ar	Ar
SSSA	Glycerol	P	P + Ar + A + O	Ar + A + O
MSSA	Glycerol	P + Ar	P + Ar + A + O	A + O

[a] SSSA = 4% Ag_2SO_4 in 95% H_2SO_4; MSSA = 20% $HgSO_4$ in 20% H_2SO_4.
[b] P = alkanes; Ar = aromatics; A = alkynes; O = alkenes.

The ideal selective-abstraction device is one which allows the class of an unknown material to be determined in a single injection. Figure 4.5(a) shows one such possible arrangement, which involves splitting the column effluent equally between various abstractors connected in parallel. Each abstractor could be designed to remove a particular class so that when, by some means, effluents were sampled in turn a record similar to that shown in Figure 4.5(b) would be obtained. The 'gap' in the record would show which abstractor had removed the eluate, and would therefore indicate its

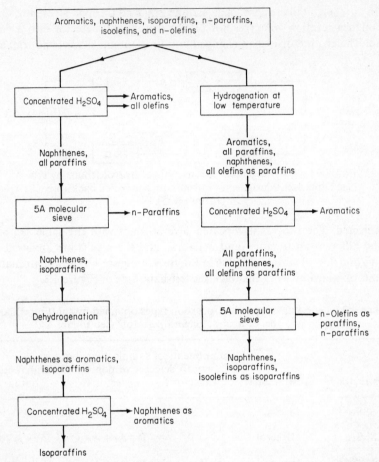

FIGURE 4.3 Flow diagram illustrating the use of selective abstraction, hydrogenation, and dehydrogenation to characterise the constituents of a complex mixture. Those components which are abstracted at each stage are shown alongside the boxes. (From ref. 1)

class. It should be emphasised that no such instrument exists, although its construction is, in principle, perfectly feasible.

Two kinds of abstractor are commonly employed. The first makes use of selective physical absorption and solution processes of the type responsible for gas–liquid and gas–solid chromatographic separations. The second kind makes use of chemical reactions which convert normally volatile compounds into involatile derivatives which the carrier gas cannot convey to the column.

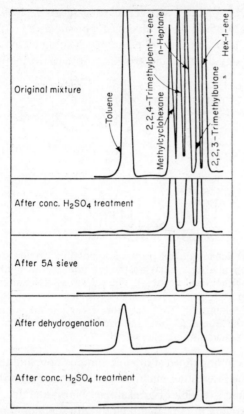

FIGURE 4.4 Chromatograms demonstrating use of the left-hand part of the identification scheme given in Figure 4.3. (From ref. 1)

4.1b Physical abstractors

A normal chromatographic column may well effectively absorb higher-boiling compounds while allowing low-boiling components to be eluted. Column materials of the diatomaceous earth type held at very low temperatures have been used to abstract traces of organic materials from air, carbon dioxide, etc.[15,16] On warming the column to normal temperatures, however, the absorbed material is recovered, and such columns are of no use for *selective* sorption. It is possible to achieve a certain amount of selectivity by the use of suitable stationary phases. For instance,[17] a coating of β,β'-thiodipropionitrile can be used to hold back benzene while allowing C_{11} saturates and olefins to be eluted. Such examples employing a liquid coating are, however, rare—a reflection of the comparatively low selectivity of conventional liquid phases. Increased selectivity is made possible by the use of a packing loaded with complexing agents such as silver nitrate in glycol for unsaturates[18] or tetrahalogenophthalate esters for aromatics,[19] and increasing attention is being paid to such columns. At the moment we are also witnessing accelerating growth of interest in the use of gas–solid

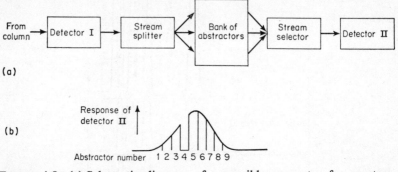

FIGURE 4.5 (a) Schematic diagram of a possible apparatus for on-stream functional group analysis. Each abstractor unit in the bank is designed to remove specific functional group(s). The stream selector directs, in sequence, the effluent from each unit to the second detector.

(b) Hypothetical output from detector II for a bank of nine abstractors, indicating that the eluate has been removed by abstractor No. 4. Vertical lines represent stream-switching transients.

columns,[20,21] including those of organic composition, and there can be little doubt that some interesting materials for selective removal of compounds will soon become available. For example, NaBr–alumina is known to be effective in removing organic compounds with functional groups (ketones, alcohols, etc.) while $AgNO_3$–alumina or CuCl–alumina is useful for the additional removal of alkenes, alkynes, and aromatics.[13] Until now, however, the only materials that have been at all widely used for selective sorption are the crystalline aluminosilicates or zeolites, which are types of 'molecular sieve'.

The ability of some materials to retain certain molecules whilst allowing others to pass is thought to be due to the existence of pores within the lattice structure of the materials which hold the retained molecules.[22] In recent years synthetic zeolites have been made in which the size of the pore can to some extent be controlled by ionic substitution.[23] The most frequently used synthetic material is the 5A sieve, which selectively retains straight-chain molecules while allowing the branched isomers to be eluted. Although this property is most commonly employed in alkane separations,[1,24,25] it can be equally valuable in analyses involving alkenes or alcohols.[11] Other interesting separations reported[11] include those of aldehydes (retained) from ketones, and of acids (retained) from esters. The 10X type sieve selectively retains aromatic materials from saturates, benzene being eluted after n-undecane.[6]

It has been reported[26] that, when acetone and propionaldehyde are injected together on to 5A molecular sieve, *both* are retained, although when injected individually only propionaldehyde is retained. This phenomenon is attributed to the occurrence of an aldol condensation catalysed by the molecular sieve, and similar reactions also occur on alumina columns.

$$CH_3-\underset{\underset{CH_3}{|}}{C}=O + CH_2-CHO \rightarrow CH_3-\underset{\underset{OH}{|}}{\overset{\overset{CH_3}{|}}{C}}-\underset{\overset{CH_3}{|}}{CH}-CHO$$

The possibility of such reactions between the components of a sample should be borne in mind when using molecular sieves for selective abstraction.

When working with high molecular weight eluates, the sieve must be maintained at a high temperature so that those molecules which are strongly (but not irreversibly) held can be eluted in a reasonable time. By raising the temperature it is possible to desorb even the persorbed n-alkanes from 5A sieves. Such desorption occurs readily near the critical temperature,[27] which for, say, C_{15} hydrocarbons is about 400 °C. In an early report of the analysis of n-alkanes from petroleum fractions by regeneration from a pre-column of 5A molecular sieve,[28] C_6-C_{10} alkanes were completely persorbed at 200 °C, so that the branched isomers could be chromatographed. The pre-column was then heated to about 400 °C to elute the n-alkanes. Similar work has been reported[25] in which at 160 °C some C_{10} and C_{11} iso-alkanes were partially absorbed by a molecular sieve pre-column. Although this process did not occur at 210 °C, n-pentane, and presumably lower n-alkanes, were not retained. A chromatogram of the abstracted higher n-alkanes was obtained by warming the pre-column to 400 °C. At such temperatures, many organic pyrolyses take place at a reasonable rate, and the method does not therefore commend itself for accurate quantitative work. It should be useful for the identification of eluates, provided that extents of pyrolytic conversion are low, as would be the case with normal flow rates and short pre-columns. More recent work[29] has employed lower temperatures around 350 °C. Pre-columns used in this work are commonly between 10 and 20 cm long.

An alternative method of regeneration, which avoids any complications due to pyrolysis, has been described.[30,31] It involves destruction of the lattice of a 5A sieve with hydrofluoric acid solution (taking care to avoid excessive heat generation) in order to liberate n-alkanes, which are apparently unaffected by the treatment. Absorption of the n-alkanes is frequently carried out off-line by refluxing with molecular sieve in benzene.[30] The sieve is then thoroughly dried and the n-alkanes are recovered with 5% hydrofluoric acid solution.

5A molecular sieves can effect the most certain separations for small alkanes, where straight and branched molecules have substantially different cross-sections, only the smaller of which is able to pass inside the zeolite framework. For larger molecules, where the size differential between straight and branched-chain isomers is not so large, it is advisable to vary the operating temperature and the extent to which the sieve is dried, in order to find the optimum conditions for separation. Although molecular sieves are proving useful, their possible applications are limited compared with those of selective chemical treatments.

4.1c Chemical abstractors

Difficulties are encountered when liquid reagents are used, principally because with reasonable flow rates it is difficult to achieve efficient absorption of the sample simply by bubbling it through the reagent. It is possible to carry out abstraction before injection (see p. 89), but the advantage of complete reaction is then often outweighed by complications arising from deposition of the reagent at the head of the column. It is for these reasons that interest has grown in the use of on-stream pre-columns containing chromatographic support material coated with reactive chemicals. Such devices expose a large surface-area of reagent to the sample, so that, provided that reaction is rapid, a short pre-column ensures complete abstraction with normal flow rates. For example, two such abstractors (see Table 4.2) are useful in the analysis of petroleum fractions: a mixture of 20% mercuric sulphate in 20% sulphuric acid completely absorbs the alkenes and alkynes, while a 4% silver sulphate in 95% sulphuric acid coating gives very efficient removal of aromatics as well as aliphatic unsaturates. On the other hand, bromine water on firebrick is ineffective.[5]

An abstraction pre-column which selectively retains alkenes and alkynes can be prepared by soaking firebrick in a mixture of $1\text{M-Hg(ClO}_4)_2$ and 2M-HClO_4 (1 : 1 by weight) and then drying at 110 °C for 1 hour. A 15 mm column of 7 mm diameter is an effective abstractor.[4] A mixture of mercuric acetate (9 g), mercuric nitrate (3 g), and ethylene glycol (20 g) distributed on 100 g of firebrick is effective in abstracting alkenes.[2]

An interesting abstractor was discovered accidentally (see Figure 4.6). Chromatograms obtained for certain samples after a particular column had been used to analyse a hydroboration reaction mixture were deficient in peaks due to alcohols definitely known to be present. This peak removal was shown to be due to a layer of boric acid on the surface of the solid

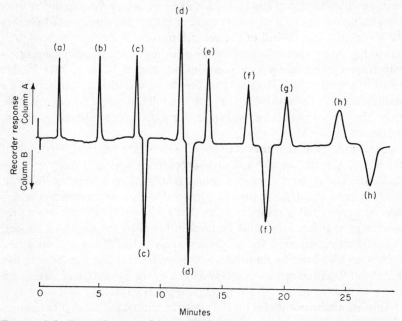

FIGURE 4.6 Recorder trace from a dual column system with its two detectors set back-to-back. Column A: 10 feet × ¼ inch, 20% Carbowax 20M; column B, as A with a post-column, 6 inch × ¼ inch, 3% boric acid. Note the time lag between the two columns. The eluates are as follows: (a) t-butyl alcohol; (b) n-butyl alcohol; (c) p-cymene; (d) linalool; (e) menthol; (f) menthyl phenyl-acetate; (g) benzyl alcohol; (h) 2-acetylpyrrole. (From ref. 7)

support, and this finding enabled efficient abstractors to be manufactured for the selective removal of alcohols from complex mixtures.[7] A column made up from 0·1 g of boric acid and 10 g of support has been used for selective removal of terpene alcohols from essential oils.[32] Traces of water produced in the process are removed by calcium hydride.

Care must be taken in the interpretation of chromatograms obtained with and without abstraction, as may be exemplified from the above work. Boric acid dehydrates rather than retains certain alcohols, notably tertiary and α-unsaturated alcohols (other than those with terminal double bonds).[7,10] If, then, the abstractor is placed *before* the column, such an alcohol will produce a peak with the retention time of the corresponding unsaturated compound. On the other hand, if the abstractor is placed *after* the analytical column, although the alcohol will still be dehydrated and proceed to the detector as an unsaturate, it will now have the retention time characteristic of the alcohol itself. Clearly, such behaviour may lead to mistaken identification, and it is advisable to confirm the validity of conclusions based on any abstraction experiment by this expedient of placing the abstractor at each end of the column in turn. A device which enables this to be done without the need for dismantling is described in connection with trapping and re-injection procedures (Chapter Nine, p. 204). In this device the trap, which is usually at the column outlet, is transferred by manipulation of three valves to the column inlet, where it acts as a sample loop. The extension of this arrangement for use with pre-columns, as above, is obvious.

An abstractor that will selectively absorb aldehydes from mixtures of aldehydes, ketones, alcohols, and hydrocarbons was also discovered by accident.[9] The material used is a commercial packing known as FFAP, a reaction product of poly(ethylene glycol) and 2-nitroterephthalic acid.[33] Reaction on FFAP is rather slow, so that for complete abstraction a pre-column several feet long must be used rather than the usual few inches. This is particularly true for aldehydes branched at the α carbon atom. FFAP has also been found to abstract epoxides.[10] It has also been found that the abstraction process is accompanied by the formation of water, and that with continued use the abstraction ability is impaired.

Another abstractor useful for the selective removal of aldehydes consists of a mull of sodium hydrogen sulphite and ethylene glycol on firebrick.[2] Several aromatic amines have also been studied as possible abstractors for aldehydes (forming Schiff bases).[10] The best reagent for specific aldehyde abstraction was found to be *o*-dianisidine (3,3'-dimethoxybenzidine). Benzidine itself efficiently abstracts aldehydes, but also abstracts most ketones and epoxides, and to a much lesser extent alcohols, esters, and ethers.

An efficient abstractor for most carboxylic acids consists of a 6-inch pre-column of a 1 : 10 mixture by weight of zinc oxide powder and column packing material. α-Substituted acids are only partially abstracted, but the remaining eluted peak tails badly, which provides a useful identification aid for such acids.[10]

The stationary phase Versamid 900 can be used for the irreversible absorption of compounds with labile halogen atoms such as alkyl halides, phenyl halides, and fatty acid α-bromo-esters.[12] Another method of chemical abstraction[34] makes use of the fact that neutral silver nitrate reacts with secondary alkyl bromides but not with primary compounds:

$$RCHBrCH_2R' + AgNO_3 \rightarrow AgBr + RCH{=}CHR' + HNO_3$$

For example, 2-bromobutane can be removed from a mixture with 1-bromo-2-methylpropane. A pre-column devised for this abstraction is made up of three distinct layers. The first is coated with the silver nitrate reagent, the second contains sulphuric acid to remove alkenes, and the third has a coating of disodium hydrogen phosphate to remove nitric acid. Experiments show that above 30 °C both secondary and tertiary bromides react, and are thus removed completely while primary isomers are unaffected. Identification by this method of the primary components of a complex mixture containing thirteen monobromo-alkanes has been reported.[34]

Maleic anhydride on silica gel can be used to remove butadiene by a Diels–Alder reaction,[3] and this technique has been extended in a novel approach to identification based upon rates of reaction, in which the extent of abstraction is studied as a function of flow rate (see p. 105).

Finally, mention should be made of the on-line chemical conversion of eluates in order to render them more easily abstractable, or to enable different functions to be distinguished in an identification scheme involving abstraction. An example of this approach has already been given in Figure 4.3, in which catalytic hydrogenation and dehydrogenation are used. By using an appropriate catalyst it is possible to hydrogenate alkenes whilst leaving aromatics unaffected.[1]

4.2 PRE-COLUMN PARTITION

Pre-column abstraction techniques using chemical reactors usually involve complete removal of certain components from the sample by means of virtually irreversible processes. Pre-column partition techniques[35–37] involve *partial* removal of an eluate by equilibration of the eluate between two immiscible solvents, followed by analysis of (usually) one of the solvents for eluate. Identification of the eluate is based upon its partition coefficient between the two solvents, or upon some equivalent parameter.

In one version of the technique,[36] the eluate is dissolved in an aliquot of solvent A (typically 5 ml) and the whole sample is analysed by GC giving an eluate peak height h_1. The same amount of eluate is then dissolved in a second aliquot of A which is shaken with an equal volume of the immiscible solvent B. Layer A is then separated and analysed as before, giving an eluate peak height h_2. Identification of the eluate is based upon its *p*-value, which is equal to h_2/h_1, and lies between 0 and 1.

There are many other possible variations of the technique. For example, instead of taking two separate aliquots of A, a single aliquot can be used, and the eluate peak-heights before and after distribution determined by analysing small samples of A. Typically 5-μl samples are taken from 5 ml of solvent.[37]

Measurement of p-values which are close to 0 or 1 is difficult by the technique as described above, since it involves the comparison of two either very similar or very dissimilar peak-heights. Such situations can be avoided by adjusting the relative volumes of A and B. For p-values close to zero, a smaller volume of B should be used, while for those close to unity larger volumes are required.

If the *detector response is linear*, and the eluate does not react with the solvents, or associate or dissociate, it is easily shown that the partition coefficient of the eluate, $K_{AB} = c_A/c_B$, is related to p by the equation

$$K_{AB} = p/(1-p)$$

Hence p-values and partition coefficients are readily interconvertible. The occurrence of such irregular phenomena as dissociation and association is indicated by differences between the p-values determined at different concentrations of eluate; they can be minimised by working at low concentrations.

Although the methods described are perfectly satisfactory, they do depend upon the accurate measurement of volume. This requirement can be eliminated by incorporating in solvent A an internal standard with a known partition coefficient, K_{AB}'. The procedure is as follows. The eluate is dissolved in an unknown volume of solvent A, and a small sample of the solution is analysed, giving peak-heights h_1 and h_1' for the eluate and internal standard, respectively. An unknown volume of solvent B is then shaken with A, and a second (small) sample of A is analysed, giving peak-heights h_2 and h_2'. It is readily shown that the partition coefficient, K_{AB}, of the eluate is then given by

$$K_{AB} = K_{AB}'[1 - (h_1'/h_2')]/[1 - (h_1/h_2)]$$

It should be noted that the volumes of A and B do not appear in this equation.

Some typical results obtained with electron-capture detection for pesticides are shown in Table 4.3. Typical errors in the p-values obtained were around 0·02, and thus many pesticides could be identified, or distinguished, with ease. Pre-column partition techniques, used in conjunction with other techniques such as retention coincidence, constitute a simple and powerful analytical tool capable of application to very low levels of eluate.

TABLE 4.3 p-Values of insecticides, determined by a single distribution between 5-ml volumes of immiscible solvents equilibrated at 25·5 °C[36]

Pesticide	Hexane–acetonitrile	Hexane–90% aq. dimethyl sulphoxide	Iso-octane–85% aq. dimethyl-formamide	Iso-octane–dimethyl-formamide
Aldrin	0·73	0·89	0·86	0·38
Carbophenothion	0·21	0·35	0·27	0·04
Gamma chlordane	0·40	0·45	0·48	0·14
p,p'-DDE	0·56	0·73	0·65	0·16
o,o'-DDT	0·45	0·53	0·42	0·10
p,p'-DDT	0·38	0·40	0·36	0·08
Dieldrin	0·33	0·45	0·46	0·12
Endosulfan I	0·39	0·55	0·52	0·16
Endosulfan II	0·13	0·09	0·14	0·06
Endrin	0·35	0·52	0·51	0·15
Heptachlor	0·55	0·77	0·73	0·21
Heptachlor epoxide	0·29	0·35	0·39	0·10
1-Hydroxychlordene	0·07	0·03	0·06	0·03
Lindane	0·12	0·09	0·14	0·05
TDE	0·17	0·08	0·15	0·04
Telodrin	0·48	0·65	0·63	0·17

A variety of solvent distribution systems, together with a selection of p-values are shown in Table 4.4. It can be seen that eluates with similar retention volumes can easily be distinguished on the basis of their p-values. Amongst the many other possible solvent systems, those involving coordinating metal ions seem to offer the most interesting possibilities.

One important feature of the technique is that it allows a check to be made on the purity of an unknown peak. Thus, if retention data or other measurements suggest that a given peak is a particular compound, pre-column partition techniques will not yield the same p-value for the peak as that for the standard material unless the peak is pure or any contaminant has the same p-value.

Chlorinated pesticides have been identified by means of the p-values of the parent compound and of the products of degradation by ultraviolet irradiation, together with the electron-capture chromatograms of these products.[38]

The partition technique is perhaps particularly useful for those eluates which can be directly trapped in a suitable solvent. The effectiveness of the general technique depends upon the behaviour of solvent/solute systems, and in this respect it has many analogies with the use of several GC columns

TABLE 4.4 p-Values at 25 °C of various compounds selected for similar retention times[37]

Compound	Retention time at 180 °C[a]	Solvent system (DMF = dimethylformamide; DMSO = dimethyl sulphoxide; % refers to aqueous solution)						
		Hexane– acetonitrile	Iso-octane– 90% DMF	Iso-octane– 90% DMSO	Heptane– 90% EtOH	Iso-octane– 80% acetone	Benzene– 80% MeOH[c]	CHCl$_3$– 60% MeOH[c]
Methyl stearate	1·55	0·95	1·0	1·0	0·87	0·98	1·0	1·0
Methyl cinnamate	1·55	0·09	0·08	0·15	0·29	0·54	0·82	1·0
Dimethyl sebacate	1·50	0·12	0·28	0·54	0·26	0·67	0·83	1·0
Diethyl sebacate	1·60	0·25	0·54	0·83	0·37	0·88	0·91	1·0
	Retention time at 100 °C[b]							
2-Methylcyclo- hexanol	2·20	0·18	[d]	0·12	0·19	0·47	[d]	0·88
3-Isomer	2·25	0·13	[d]	0·06	0·11	0·26	[d]	0·83
4-Isomer	2·30	0·08	[d]	0·05	0·18	0·33	[d]	0·84

[a] Column used, 5% (w/w) diethylene glycol succinate on 60–80 mesh Diaport S.
[b] Column used, 5% (w/w) silicone oil 550 on 100–110 mesh Anakrom ABS.
[c] p-Values based on analysis of the lower phase.
[d] Solvent interferes with determination.

in identification by retention. However, it has the enormous advantage over this procedure of being experimentally much easier to effect, and it is also more versatile. The changing of a solvent–solvent system is very easy compared with the changing of GC columns; even if some form of multi-column system is employed with the latter, it is not highly practicable for more than a few columns. Also, the requirements of solvents in the pre-column partition technique are not as stringent as those in gas–liquid chromatography: the solvent must not be too volatile, must not elute in the region of the unknown eluate, and must not interfere markedly with the operation of the detector. Apart from these broad requirements there is a wide choice of solvents available.

REFERENCES

1. R. Rowan, 'Identification of hydrocarbon peaks in gas chromatography by sequential application of class reactions', *Analyt. Chem.*, **33**, 658 (1961).
2. J. A. Kerr and A. F. Trotman-Dickenson, 'Absorbents for aldehydes and olefins', *Nature*, **182**, 466 (1958).
3. J. Janak and J. Novak, 'Chromatographic semimicroanalysis of gases. XIV. Direct determination of individual gaseous paraffins and olefins in 1,3-butadiene', *Chem. Listy*, **51**, 1832 (1957) (*Chem. Abs.*, **52**, 1860d).
4. D. M. Coulson, 'Hydrocarbon compound-type analysis of automobile exhaust gases by mass spectrometry', *Analyt. Chem.*, **31**, 906 (1959).
5. W. B. Innes, W. E. Bambrick, and A. J. Andreatch, 'Hydrocarbon gas analysis using differential chemical absorption and flame ionization detectors', *Analyt. Chem.*, **35**, 1198 (1963).
6. J. V. Brunnock and L. A. Luke, 'Determination of hydrocarbon type composition of petroleum distillates boiling up to 185 °C using type X molecular sieves', *Analyt. Chem.*, **41**, 1126 (1969).
7. R. M. Ikeda, D. E. Simmons, and J. D. Grossman, 'Removal of alcohols from complex mixtures during gas chromatography', *Analyt. Chem.*, **36**, 2188 (1964).
8. T. Sato, N. Shimliki, and N. Mikami, *Bunseki Kagaku*, **14**, 223 (1965).
9. R. R. Allen, 'Selective removal of aldehydes from complex mixtures during gas chromatography', *Analyt. Chem.*, **38**, 1287 (1966).
10. B. A. Bierl, M. Beroza, and W. T. Ashton, 'Reaction loops for reaction gas chromatography. Subtraction of alcohols, aldehydes, ketones, epoxides, acids, and carbon-skeleton chromatography of polar compounds', *Mikrochim. Acta*, 637 (1969).
11. N. Brenner, E. Cieplinski, L. S. Ettre, and V. J. Coates, 'Molecular sieves as subtractors in gas chromatographic analysis. II. Selective adsorptivity with respect to different homologous series', *J. Chromatog.*, **3**, 230 (1960).
12. M. Rogozinski, 'Subtraction gas chromatography of labile halogen compounds', *J. Gas Chromatog.*, **2**, 163 (1964).
13. C. S. G. Phillips and C. G. Scott, 'Modified solids for gas–solid chromatography', *Progress in Gas Chromatography* (ed. J. H. Purnell), Interscience, New York (1968), p. 121.

14. N. Brenner and V. J. Coates, 'Molecular sieves as subtractors in gas chromatographic analysis', *Nature*, **181**, 1401 (1958).
15. J. H. Williams, 'Gas chromatographic techniques for the identification of low concentrations of atmospheric pollutants', *Analyt. Chem.*, **37**, 1723 (1965).
16. P. S. Farrington, R. L. Pecsok, R. L. Meeker, and T. J. Olson, 'Detection of trace constituents by gas chromatography', *Analyt. Chem.*, **31**, 1512 (1959).
17. R. L. Martin, 'Determination of hydrocarbon-types in gasoline by gas chromatography', *Analyt. Chem.*, **34**, 896 (1962).
18. M. A. Muhs and F. T. Weiss, 'Determination of equilibrium constants of silver-olefin complexes using gas chromatography', *J. Amer. Chem. Soc.*, **84**, 4697 (1962).
19. S. H. Langer, C. Zahn, and G. Pantazoplos, 'Selective gas–liquid chromatographic separation of aromatic compounds with tetrahalophthalate esters', *J. Chromatog.*, **3**, 154 (1960).
20. D. H. Everett, 'The interaction of gases and vapours with solids', in *Gas Chromatography 1964* (ed. A. Goldup), Institute of Petroleum, London (1965), p. 219.
21. S. Dal Nogare, 'New developments in gas chromatography', *Analyt. Chem.*, **37**, 1450 (1965).
22. R. M. Barrer, 'Molecular-sieve action of solids', *Quart. Rev.* (London), **3**, 293 (1949).
23. D. W. Breck, 'Crystalline molecular sieves', *J. Chem. Educ.*, **41**, 678 (1964).
24. P. A. Schenck and E. Eisma, 'Determination of *n*-alkanes in mixtures of saturated hydrocarbons by means of gas chromatography', *Nature*, **199**, 170 (1963).
25. D. K. Albert, 'Determination of C_5 to C_{11} *n*-paraffins and hydrocarbon types in gasoline by gas chromatography', *Analyt. Chem.*, **35**, 1918 (1963).
26. L. S. Ettre and N. Brenner, 'Molecular sieves as subtractors in gas chromatographic analysis. III. The secondary effect of the molecular sieve trap column', *J. Chromatog.*, **3**, 235 (1960).
27. W. P. Ballard, S. P. Dickens, and B. F. Smith, *U.S. Pat.* 2,818,455 (1957).
28. F. T. Eggertsen and S. Groennings, 'Determination of small amounts of *n*-paraffins by molecular sieve-gas chromatography', *Analyt. Chem.*, **33**, 1147 (1961).
29. G. C. Blytas and D. L. Peterson, 'Determination of kerosene range *n*-paraffins by a molecular sieve gas–liquid chromatography method', *Analyt. Chem.*, **39**, 1434 (1967).
30. J. V. Brunnock, 'Separation and distribution of normal paraffins from petroleum heavy distillates by molecular sieve adsorption and gas chromatography', *Analyt. Chem.*, **38**, 1648 (1966).
31. H. S. Knight, 'Gas chromatographic determination of normal paraffins in kerosene', *Analyt. Chem.*, **39**, 1452 (1967).
32. F. W. Hefendehl, 'Elimination of terpene alcohols by reaction gas chromatography with boric acid', *Naturwiss.*, **51**, 138 (1964) (*Chem. Abs.*, **60**, 15677c).
33. W. K. Lee and R. M. Bethea, 'Gas chromatographic analysis of aliphatic oxygenated compounds in the presence of water', *J. Gas Chromatog.*, **6**, 582 (1968).

34. W. E. Harris and W. H. McFaddy, 'Selective reactivity in gas–liquid chromatography: determination of 2-bromobutane and 1-bromo-2-methylpropane', *Analyt. Chem.*, **31**, 114 (1959).
35. R. Suffis and D. E. Dean, 'Identification of alcohol peaks in gas chromatography by a non-aqueous extraction technique', *Analyt. Chem.*, **34**, 480 (1962).
36. M. Beroza and M. C. Bowman, 'Identification of pesticides at nanogram level from extraction *p*-values', *Analyt. Chem.*, **37**, 291 (1965).
37. M. C. Bowman and M. Beroza, 'Identification of compounds by extraction *p*-values using gas chromatography', *Analyt. Chem.*, **38**, 1544 (1966).
38. K. A. Banks and D. D. Bills, 'Gas chromatographic identification of chlorinated insecticides based on their U.V. degradation', *J. Chromatog.*, **33**, 450 (1968).

CHAPTER FIVE

TECHNIQUES OF IDENTIFICATION INVOLVING CHEMICAL MODIFICATION OF THE SAMPLE

Chemical processes are involved in many identification techniques in GC. This chapter is concerned with those 'chemical' methods which involve relatively simple reactions and are often firmly based on preparative and analytical chemistry.

Some authors have called this general area of GC 'reaction gas chromatography', which includes selective abstraction and degradative methods. Strictly, this term only applies to reactions which occur in a single integrated operation somewhere within the closed GC system, such as those of on-stream reactors. However, there are many other techniques employing the same basic approach which need not necessarily be integrated, and these methods are included in the contents of this chapter.

Chemical modification in GC may take place prior to injection, after elution, or even on the partitioning column. Within the scope of this chapter, identification then follows from one of the following.

(a) The retention volume of the unknown and that of a derivative formed in the reaction.

(b) Identification of the derivative by some other means.

(c) Physical changes (e.g. colour) accompanying the reaction.

(d) The rate of reaction in the system.

5.1 FUNCTIONAL GROUP ANALYSIS

Functional-group analysis in GC can be carried out before or after the column. In the latter technique operations are performed on separate peaks while in the former the whole sample is involved.

5.1a Post-column

Methods of conventional qualitative analysis rely heavily upon the abilities of specific functional groups to cause a colour change, precipitation, or other obvious event with the appropriate reagent. Such tests can be adapted for use with GC, thereby enabling individual peaks to be assigned to their class of compound.[1,2] One way of doing this is to use a multi-way stream splitter at the column outlet, and to bubble the branches of effluent through reagents contained in phials. After any colour change or other reaction has taken place with a particular reagent, the appropriate phial is replaced with a fresh one. This operation takes only a few seconds, and so virtually continuous monitoring of all but the fastest chromatograms is possible. Alternatively, the effluent can be switched between two banks of phials, the used phials of the 'fallow' bank being replaced ready for use again.

Walsh and Merritt,[1] using a five-way stream splitter in this way, were able to classify peaks containing any one of eleven functional groups in (at most) two runs with the aid of a dozen reagents. The simplicity and wide applicability of their method make it worthwhile to outline the classification scheme (see Table 5.1). One disadvantage of the method as it stands is that fairly high peak-loads are required (~ 50 μg with a five-way splitter).

Retention volume/carbon number correlation plots (see Chapter Three) are really only effective when peaks have been classified according to their functional group. In addition, such plots often enable ambiguities to be resolved. For example, if a positive test were obtained with 2,4-dinitrophenylhydrazine the eluate could be an aldehyde or a ketone and, within these classes, a straight-chain, or branched-chain, or aromatic compound, etc. Reference to the appropriate correlation plots would generally show which it is since the carbon number of any real compound must be integral. Similarly, bifunctional compounds may be distinguished from peaks due to a mixture of co-eluted compounds. However, this approach may be misleading if full account is not taken of all the relevant retention data.

In an ingenious variation of the post-column method use is made of the capability of GC to separate *in space* the components of a mixture.[3] A horizontally advancing strip, of the type used in thin-layer chromatography, is positioned under the column outlet of the chromatograph. The strip moves at the same rate as the detector recorder chart, so the position of a component on the strip can be pinpointed from the position of the corresponding peak on the chart. The strip is wetted with reagents of the type shown in Table 5.1, thereby enabling those peaks which belong to a given class to be identified (see Figure 5.1).

It is sometimes found that some very large peaks correspond to very light colour spots when developed on the strip. This situation can be attributed to the overlap of a minor component containing the relevant functional group by a major component without that group. Few other techniques enable overlapped peaks to be detected so simply.

The moving-strip system is not well suited for use with more than one reagent per run, although longitudinally divided strips can be used with

D

TABLE 5.1 Post-column qualitative analysis[1]

No.	Reagents	Colour after reaction	Functional group indicated	Minimum detectable amount of eluate (μg)	Other reagents to help classification
1	Ceric nitrate	Yellow-amber	Primary, secondary, or tertiary alcohol	100	8
2	2,4-DNPH	Yellow ppt.	Aldehyde or ketone	20	9
3	Ferric hydroxamate in propylene glycol	Red	Ester or nitrile	40	10
4	Sodium nitroprusside	Red	Mercaptan, sulphide, disulphide, or primary amine	50	11, 12
		Blue	Secondary amine		7
5	Formaldehyde and sulphuric acid	Wine-red	Aromatic nuclei or unsaturated aliphatics	50	Distinguish by retention data
6	Alcoholic silver nitrate	White ppt.	Alkyl halide	20	13
7	Hinsberg's	Orange	Primary or secondary amine	40	4
8	Nitrochromic acid	Yellow to blue-grey	Primary or secondary alcohol	20	1
9	Schiff's	Pink	Aldehyde	50	2
10	Ferric hydroxamate	Red	Ester	40	3
11	Isatin	Green	Mercaptan or disulphide	100	4, 12
12	Lead acetate	Yellow ppt.	Mercaptan	100	4, 11
13	Mercurous nitrate	Yellow to orange ppt.	Iodide	20	
		White or grey ppt.	Bromide or chloride		6

each half wetted with a different reagent, the effluent being applied along the dividing line.

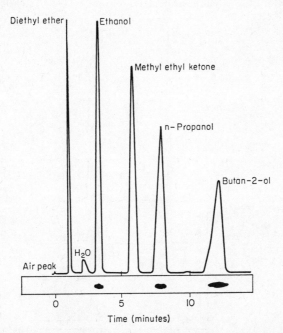

FIGURE 5.1 Post-column functional group analysis. Gas chromatogram of a $2 \mu l$ sample of a mixture of oxygenated compounds (top) compared with corresponding paper strip wetted with the reagent nitrochromic acid and moved past the column outlet at the same rate as the detector chart. (From ref. 3)

5.1b Pre-column

Comparison of the chromatograms obtained with and without pre-column chemical modification often enables the functional groups of eluates to be determined. One promising technique, which is suitable for vapours at concentrations from about 10^{-5} to 10^{-8} g/ml, uses a syringe (typically 5 ml volume) as the reaction chamber.[4] The walls of the syringe, which are of ground glass, are wetted with reagent. The sample vapour is then introduced and reaction is allowed to proceed for (typically) a few minutes. The entire reaction mixture is then injected into the chromatograph, and the resulting chromatogram is compared with that of the untreated sample. Interference from the reagent can, if necessary, be avoided by transferring the chemically modified vapour to another syringe before injection.

A large number of conventional reagents are suitable for use with this technique (see Table 5.2). Figures 5.2 and 5.3 show examples of selective removal and peak shifting, respectively.

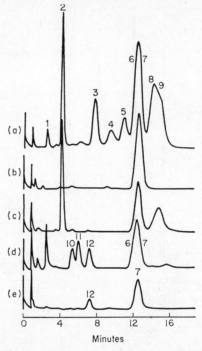

FIGURE 5.2   Chromatograms obtained from a multi-component vapour mixture after treatment with various reagents in a syringe.
(a) Untreated sample; (b) concentrated H_2SO_4; (c) sodium metal; (d) following (c), ozone (note ozonide decomposition products 6, 10, 11, and 12); (e) following (c) and (d), hydroxylamine hydrochloride.
Key to peaks: 1, acetaldehyde; 2, ethyl vinyl ether; 3, methyl acetate; 4, methanol; 5, methacraldehyde; 6, n-butyraldehyde; 7, n-heptane; 8, butan-2-one; 9, hept-3-ene; 10, propionaldehyde; 11, formaldehyde; 12, ethyl formate. (From ref. 4)

Caution must be exercised with the use of some reagents. For example, bromine not only removes unsaturated materials, but also to a lesser extent affects other compounds, particularly alcohols and aldehydes. Similarly, the removal of carbonyls with acid hydroxylamine also causes some loss of alkyl sulphides and alcohols.[5] In addition, the removal of some components by their solution in the reagent may produce anomalous results. The effects of common reagents on a number of different functional groups are given

TABLE 5.2 Class reactions of some common reagents[a]

Reagent[a]	Details of preparation	Class, and effect of reagent[d]							
		ROH	RCHO	RCOR'	RCOOR'	ROR'	R=R	Ar	RH
Na	Metal slice	A	A	A	A	E	E	E	E
H_2SO_4	Concentrated	A	A	A	A	A	A	E	F
7 : 3 H_2SO_4	7 : 3 (v/v) of conc. H_2SO_4 and water	A	C	A	C	C	E	F	F
H_2	H_2 gas and a few mg of PtO_2	A	A	E	F	E	B	B	F
HI	Prepared from 2 ml 90—95% H_3PO_4 warmed with a few mg of KI	A	A	A	A	A	E	E	E
Br_2	Freshly prepared saturated solution	D	D	E	E	E	B	F	F
NH_2OH	4 g of hydrochloride in 50 ml water	C	A	A	F	F	F	F	F
$NaBH_4$	1 g in 2 ml of water	C	B	B	E	E	F	F	F
$KMnO_4$	Saturated solution	D	A	E	E	E	C	F	F
$NaNO_2$	Equal amounts of $NaNO_2$ solution (2·5 g in 50 ml) and $1N$-H_2SO_4	B	E	E	E	F	F	F	F
$(CH_3CO)_2CO$[b]	5 ml acetic anhydride and 2 drops conc. H_2SO_4	B	D	C	E	E	F	E	F
NaOH	2·5 g NaOH in 50 ml water	C	C	E	B	F	F	F	F
O_3[c]	Ozone in oxygen	D	E	E	E	E	B	F	E

[a] Typical amounts of reagent used were 5–10 μl in syringes of between 2 and 10 ml.

[b] After reaction with the sample the mixture is neutralised with $NaHCO_3$.

[c] Ozonides are reduced to carbonyl compounds using $NaAsO_2$ solution (10% w/w) or triphenylphosphine.

[d] A, Removal; B, removal and formation of new products; C, severe decrease; D, severe decrease and formation of new products; E, slight decrease; F, no effect.

in Table 5.2: the recommended order in which to use the reagents on an uncharacterised sample is from top to bottom in this table.

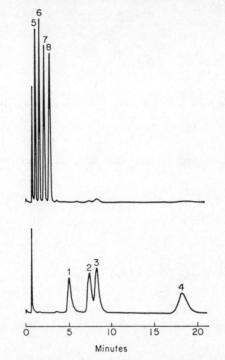

FIGURE 5.3   Peak-shifting of alcohols (bottom) in the formation of nitrites (top) by treatment with nitrous acid in a syringe prior to injection. Both chromatograms were obtained on the same column (15% Ucon non-polar LB 1715 on Chromosorb W).
Key to peaks: 1, CH_3OH; 2, C_2H_5OH; 3, isopropyl alcohol; 4, n-propyl alcohol; 5, 6, 7, and 8, corresponding nitrites. (From ref. 4)

Both pre-column and post-column methods of functional-group analysis are relatively simple and quick, and enable the analyst to sketch in the types of compound present in an uncharacterised sample with the aid of reagents which are available in almost any laboratory. They are a good starting point from which to single out which of the more sophisticated methods of identification might be most useful.

5.2 OZONOLYSIS

Determination of the position of carbon–carbon double bonds by means of ozonolysis is well known in conventional organic chemistry. With the aid of GC the technique can be extended to very much smaller quantities of sample than is usual.[6,7]

Ozonides decompose thermally to form acidic and aldehydic fragments. In some methods this process is made to occur as the ozonide is injected through a heated injection port.[7,8] Identification of the major products then frequently enables the groups attached to the double bond to be determined. Acidic and aldehydic products of the process may be distinguished by using a short pre-column of zinc oxide granules for selective removal of the acids.

Pyrolysis of ozonides in the presence of a hydrogenation catalyst (palladium on charcoal) yields only aldehydic products.[8] Similar results are obtained by hydrolysis in the presence of zinc or by reduction with triphenylphosphine.

$$\underset{\underset{O}{\diagdown\diagup}}{\overset{O-O}{\overset{\diagup\quad\diagdown}{RCH\quad CHR'}}} + Ph_3P \rightarrow RCHO + R'CHO + Ph_3PO$$

Triphenylphosphine reduction has been used in an attractive technique requiring, in some cases, as little as 1 μg of sample.[6] The ozonisation apparatus is shown in Figure 5.4. Ozone, produced in a silent discharge, is passed into a solution of the sample in carbon disulphide or pentyl acetate cooled by solid carbon dioxide in the tube F. The effluent from this tube is passed into an indicator solution containing potassium iodide, sulphuric acid, and starch until the characteristic blue colour is produced; the ozone supply is then stopped. Ozonisation is generally complete within 10—15 seconds, after which the solution is removed from the cold-bath, 1 mg of triphenylphosphine is added, and the mixture is allowed to reach room temperature. An aliquot is then injected into the chromatograph, and the aldehydic components of the chromatogram are identified. In this context it is worth mentioning again that aldehydes are selectively abstracted by the liquid phase FFAP[9] (see p. 78).

This method of ozonolysis is both simple and quick, and does not require expensive equipment. It should be particularly useful in the determination of the position of double bonds in the side-chains of large molecules, since only the aldehydes which result from the side-chain will be detected under normal chromatographic conditions, and they will generally be very simple.

The products of ozonolysis of triple bonds in methanol solution are mainly methyl esters, whereas double bonds as usual yield mainly aldehydes. Chromatograms of the reaction products have been used to characterise straight-chain compounds containing double and/or triple bonds. This technique has also proved useful for locating the double bonds and hydroxy group in α-hydroxy conjugated dienes.[10]

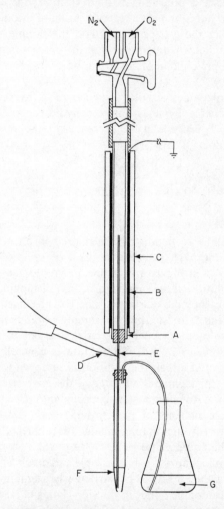

FIGURE 5.4   Apparatus for ozonisation of microgram quantities of sample. Ozone passes from the micro-ozoniser (A, glass; B, aluminium foil; C, rubber; D, high-voltage source such as vacuum tester; E, small-bore tubing) into the sample F. The outflow is passed into the indicator solution G ($KI–H_2SO_4$–starch) until a blue colour indicates an excess of ozone and therefore completion of reaction. (From ref. 6)

5.3 GAS CHROMATOGRAPHY OF DERIVATIVES: PEAK-SHIFT TECHNIQUES

As in conventional organic analysis, it is frequently convenient in GC to work with derivatives of an eluate. These may be used for a number of reasons. In the first place, the volatility of a compound may be increased so that it can be chromatographed in the gas phase. In this way many 'involatile' materials such as amino-acids, carbohydrates, and other substances, generally involving extensive hydrogen-bonding in the condensed phase, have been analysed by GC.

Secondly, a derivative is sometimes prepared to increase the sensitivity of detection of an eluate by substituting it with halogenated groups suitable for detection by electron capture (EC). For example, steroidal heptafluoro-butyrates have a higher EC sensitivity than the parent materials. They are also more volatile.

Thirdly, in some cases, selective pre-column abstraction of functional groups can be carried out by forming suitable derivatives. For example, ketones and aldehydes can be precipitated from solution as 2,4-dinitro-phenylhydrazones. The parent compounds can be regenerated for chromatography by heating with α-ketoglutaric acid in the injection port of the chromatograph. Typically, regeneration at 250 °C takes about 10 seconds.[11,12] Alternatively, the derivatives themselves can be chromato-graphed,[13] an approach which has also been used for aromatic amines.[14]

The fourth, and in the present context the most important, reason for using a derivative is to provide a second piece of information, namely, the retention volume of the derivative, for use in identification of the eluate from retention data. Thereby, the certainty of identification is increased in a way which is analogous to the use of two columns in retention-coincidence studies, and also to the determination of melting points of derivatives in conventional organic analysis. This 'peak shift' method of identification, specifically advanced by Langer and Pantages[15] for the trimethylsilyl esters of hydroxy-compounds, can be applied to any material which is sufficiently volatile itself to be eluted from a GC column, and for which a volatile derivative is available. Figure 5.3 shows peak-shifting of alcohols as nitrite esters.

Since it is necessary to know the type of material being analysed before a suitable derivative can be prepared, peak-shift methods are inevitably restricted to those samples which are reasonably well characterised, just as melting-point methods are used to pinpoint the exact identity of a compound after its functional groups have been determined. Thus peak-shift methods might best be employed after a reasonable amount of preliminary sample characterisation has been carried out. The functional group tests described above (pp. 86—92) are ideal for this purpose in many cases.

Not all derivatives are equally suitable for GC analysis. They should be stable at the elution temperature, and should give sharp symmetrical peaks. It is desirable that they can be easily prepared by a method which leads unambiguously to a single derivative by the use of a reagent which does not interfere with subsequent analysis, or alternatively one which is easily removed by a process such as vacuum distillation. It is a great advantage if the derivative can be formed on-column, thereby enabling the eluate to be kept in the gas phase.[16]

In peak-shift methods it is essential that the derivative and its parent compound have significantly different retention volumes. Hence, if a non-polar column is used, the derivative must have a significantly higher molecular weight than its precursor. This can be seen by reference to Figure 2.3. It should be noted that if the group additivity characteristics of retention were accurately obeyed (p. 53), peak-shift methods would be of little value, since the retention volumes of all derivatives would be in the same fixed ratio to those of the parent compounds. Fortunately such 'rules' are generally only approximately true, especially for the lower members of homologous series.

5.3a   Experimental methods

Peak-shift techniques are best performed on separated peaks. In general, the eluate is first chromatographed by itself, and its retention volume noted. A second sample is then treated to form the derivative. This process may take place external to the GC apparatus, or in a bypass chamber from which reaction products are injected as a plug.

Other experimental variants include forming the derivative on-column by phased injection of the reagent and sample. For example, alcohols, phenols, and primary and secondary amines can be esterified by injecting an anhydride a few seconds after the sample.[16] It is essential that retention times are of the order of tens of minutes so that the time between sample injection and completion of reaction does not become critical. Two new peaks normally result if the anhydride is in excess, but when two esterifiable groups are present it is possible to obtain four peaks. By using simultaneous injection of two different anhydrides, a further aid to identification is achieved by the formation of two product esters from a single functional group (see Figure 5.5). Alkaloids and steroids[41,66] can be characterised by this technique, and, using acetic and propionic anhydrides, it has been employed to assist in the identification of narcotic analgesics.[17] On-column esterification, which can readily be applied to very large molecules (e.g.

valerial sesquiterpenoids[30]), should find widespread biomedical application. On-column esterification with methanolic hydrochloric acid has recently allowed gas chromatographic analysis of some dealkyl metabolites of organophosphorus insecticides.[18]

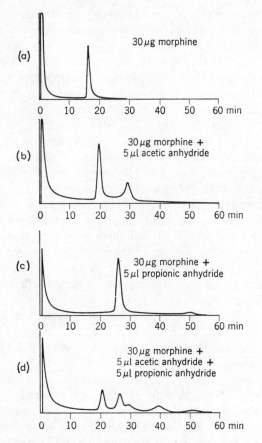

FIGURE 5.5 Chromatograms illustrating peak shift after precolumn reaction. The anhydrides are added in large excess. The new peak near 40 min in (d) is probably due to a mixed ester (morphine contains two hydroxy groups). (From ref. 16)

Frequently, of course, retention times are so short, or reaction times so long, that this approach would be useless. For example, it has been found that trimethylsilyl ethers cannot be prepared by this technique unless the carrier gas supply is stopped for about 5 minutes after injection of the reagent hexamethyldisilazane. In such circumstances it is obviously simpler to carry out reaction before injection, as in the pre-column syringe technique[4] described above (p. 89).

5.3b   Derivatives of alcohols, phenols, and other hydroxy-compounds

The usual derivatives of hydroxy-compounds are the trimethylsilyl ethers, which are rapidly formed at room temperature by reaction with hexamethyldisilazane.[15,19] The reaction, which is normally complete within a few minutes, is catalysed by trimethylchlorosilane, and is frequently carried out in pyridine solution. All materials must be scrupulously dry since both reagents and the derivatives are readily hydrolysed. A recent book[19] on silylation gives retention data for many derivatives.

Trimethylsilyl ethers frequently, but not invariably, have smaller retention volumes than their parent compounds. They are well suited to peak-shift measurements since they are easily prepared from a wide variety of compounds, including aliphatic alcohols,[15,20] glycols,[21] polyhydric alcohols,[22] carboxylic acids,[23] steroids,[24] and carbohydrates and related polyhydroxy-compounds.[25,26] The trimethylsilyl ethers of carbohydrates allow the resolution of anomeric pairs as well as configurational isomers for eluates with molecular weights as high as 1500. For an exhaustive review of the use of derivatives of carbohydrates in GC, the reader is referred to the literature.[27] A study has been made of the liquid phases and solid supports suitable for the trimethylsilyl derivatives of monosaccharides.[28]

Chloromethyldimethylsilyl ethers[29] have also found use in GC, particularly for steroids, in which instance they are more stable than the trimethylsilyl ethers. These halogenated derivatives are also useful for electron capture detection (p. 169).

Acid ester derivatives—particularly acetates—of hydroxy-compounds are frequently used. They have the advantage over trimethylsilyl ethers that they can be prepared from the corresponding anhydride without the need for very dry reaction conditions. On the other hand, acetylation is sometimes slower than silylation, although phenols and related compounds are rapidly acetylated. Acetates are generally more stable than trimethylsilyl ethers, particularly towards hydrolysis. Also, as discussed on p. 96, they can be prepared on-column by injection of the anhydride a few seconds after the hydroxy-compound, while trimethylsilyl ethers cannot be prepared in this way.

Halogenated esters are sometimes used owing to their high volatility. For example, chloroacetates are useful derivatives of phenols, and can be rapidly formed by shaking them with acetic anhydride in alkaline benzene solution.[31]

Various methods of analysing aqueous alcohols with the aid of a reactive pre-column have been devised. Conversion into nitrite esters (cf. Figure 5.3) was effected by passing the acidified solution through a reactor containing sodium nitrite on a solid support.[32,33] In another method, alcohols were dehydrated to the corresponding olefins by passing them through a bed of phosphorus pentoxide on silica at 300 °C. Water was removed in a calcium hydride column prior to separation of the derivative olefins.

5.3c Reactions of carboxylic acids and esters

Carboxylic acids are generally chromatographed as their methyl esters since the free acids have poor chromatographic properties.[34,35] Hence, peak-shifts between the methyl ester and *another* derivative are probably of greatest interest. Trimethylsilyl ethers are readily prepared,[23] and are particularly useful for hydroxy-acids since *both* functional groups are silylated.

Methyl and higher esters are conveniently and rapidly prepared by pyrolysis of the corresponding tetraalkylammonium salt. This process can be carried out in an injection port packed with glass wool and maintained at about 370 °C.[36] Possible interference from product alkylamine can be avoided by using a short abstraction pre-column of ion-exchange resin.[37]

Ethyl esters can be formed by 'flash exchange' with potassium ethyl sulphate. The free acid, or its potassium or sodium salt, is mixed with the reagent in aqueous solution, and an aliquot is drawn into a hypodermic needle containing diatomaceous earth. The water is evaporated at 100 °C, and the needle is then inserted into a heated injection port (temperature ~275 °C), thereby enabling the ethyl ester to be formed and injected in a concerted operation.[38,39]

A useful aid to the identification of esters involves saponification in a pre-column of wet potassium hydroxide. The resulting acids are retained, and only the alcohols proceed to the GC column.[40]

An example of the application of peak-shift techniques to peak identification is illustrated in Figure 5.6, which shows the retention data for methyl, trifluoroacetyl, and trimethylsilyl derivatives of bile acids from human faeces.[41] There is sufficient differentiation between the retention of the derivatives to enable identification of the parent materials. Thus, although peaks 1 and 2 are closely eluted as methyl esters, they are well separated as trifluoroacetyl and particularly trimethylsilyl derivatives.

Peak-shifting by oxidation to the ketocholanate was also used in this work. All the 5β-acids formed the same ketocholanate with retention around 0·6 of that formed from the 5α-acid.[41] The two types of acid could therefore be readily distinguished.

5.3d   Derivatives of amines and amino-acids

The analysis by GC of amino-acids has attracted a great deal of interest because of its important biochemical applications. The acids themselves are not sufficiently volatile to be chromatographed in the gas phase, and attention has therefore turned to the preparation of volatile derivatives. An early approach to the problem was reduction of the α-amino-acid to the next lower aldehyde by reaction with ninhydrin.[42] This technique was

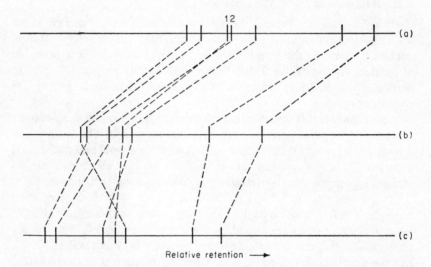

FIGURE 5.6   Retention of derivatives of bile acids relative to methyl deoxycholate: (a) methyl ester; (b) trifluoroacetate; (c) trimethylsilyl ether. Peak-shifting is particularly marked between derivatives (b) and (c). (Data from ref. 41)

adapted so that it could be carried out on a pre-column.[43,44] The amino-acid solution in hydrogen carrier gas was passed into a micro-reactor at 130 °C containing 30% (w/w) ninhydrin distributed on diatomaceous earth.

More recently a wide variety of derivatives has been prepared and examined, and an exhaustive review of this work has been published.[45] At the present time it appears that the n-alkyl esters of the N-trifluoroacetyl derivatives of the amino-acids are the most satisfactory.[46-48] They are simply prepared by reaction of the alkyl ester with trifluoroacetic anhydride, and can be purified by silica gel column chromatography.[49] Amines and ethanolamines are also readily converted into mono-N-trifluoroacetyl derivatives by reaction with trifluoroacetic anhydride.[50-52] Prolonged reaction may lead to substitution of both hydrogen atoms of the amine.

Amines may also be chromatographed as their trimethylsilyl ethers,[53] as enamines, and as a variety of amide derivatives.[54] A simple and convenient procedure is acetylation, which can be carried out on the amine hydrochloride (say 10 mg) with acetic anhydride (0·1 ml) in acetonitrile (0·3 ml) and pyridine (0·1 ml). The resulting acetyl derivatives may be injected on to the column without isolation.

5.4. HYDROGENATION AND DEHYDROGENATION

Many of the catalytic processes devised for hydrogenation and dehydrogenation involve particulate beds, and they are therefore ideally suited to the requirements of on-stream reaction GC. Thus it is relatively easy to convert alkenes into alkanes, naphthenes into aromatics, etc., using one of a large number of available catalysts. Hydrogenation catalysts usually consist of freshly reduced finely divided metals (Ni, Pt, or Pd) distributed on a solid support and operated up to about 200 °C. In solution, Adams platinum oxide catalyst is generally used. Dehydrogenation, which is less frequently used and is not normally such a 'clean' process, can be effected with a variety of catalysts, including chromia or chromia–alumina operated at about 400 °C.

From the point of view of identification these processes are useful in two ways. First, they enable materials to be labelled 'saturated' or 'unsaturated',[55] and secondly, they provide two bits of information for each peak, i.e. its retention volume before and after reaction.[56] The gain or loss of a molecule of hydrogen makes very little difference to the molecular weight of an eluate, and may scarcely change its boiling point. Therefore, if any significant change in retention volume is to be produced by reaction, it is necessary to use a polar column which retains molecules according to their degree of saturation. In comparison, many of the derivatives described in Section 5.3 involve large changes in molecular weight.

5.4a Experimental methods

A convenient way of determining whether an eluate has gained or lost hydrogen, or is unchanged, is to monitor the concentration of hydrogen leaving the reactor. Klesment[57] achieved this by using an inert carrier gas containing a small proportion of hydrogen. The effluent from the GC column passed through a reactor (5% Pt on kieselguhr), after which the products of reaction were absorbed and the hydrogen concentration determined with a katharometer.

Another interesting technique for use with hydrogen-containing carrier gas employs two katharometers connected in series, one of which has

CARL A. RUDISILL LIBRARY
LENOIR RHYNE COLLEGE

filaments (or thermistors) coated with a suitable hydrogenation catalyst (e.g. Pd). Materials which are not hydrogenated yield equivalent signals from both detectors, whereas hydrogenatable substances do not since the heat of hydrogenation modifies the normal thermal response.[68]

A similar but more sophisticated device has been described by Littlewood and Wiseman.[58] Electrolytically generated hydrogen is mixed with the column effluent and passed through a catalytic reactor. The modified effluent then proceeds to an electrochemical cell which responds only to hydrogen and not to the products of reaction. The output from this cell is linked to the electrolytic hydrogen generator in a feedback loop which compensates for any loss or gain of hydrogen due to hydrogenation/ dehydrogenation reactions. Thus, the current through the generator traces, in the form of peaks (negative and positive), the elution of those eluates

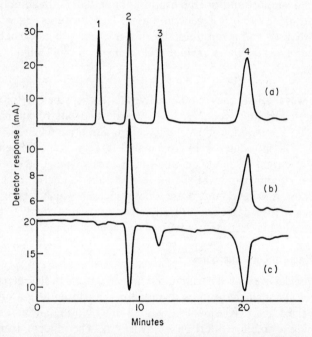

FIGURE 5.7 Chromatograms of a mixture of hydrocarbons obtained with 'servomechanism' detectors (see text) which respond to (a) combustion, (b) hydrogenation, and (c) dehydrogenation of eluates.

Key to peaks: 1, 2,2-dimethylbutane; 2, hex-1-ene; 3, 2,4-dimethylpentane; 4, hept-1-ene. (From ref. 58)

which lose or gain hydrogen. A similar device controlled by oxygen concentration can be used to distinguish those eluates which burn. Figure 5.7 shows chromatograms obtained with each device for a mixture of alkanes and alkenes.

5.5 CARBON SKELETON CHROMATOGRAPHY

In this technique,[59] compounds are stripped of their functional groups, and hydrogenated or dehydrogenated (see previous section) by passing them in a stream of hydrogen carrier gas through a catalytic pre-column held at an elevated temperature. Sometimes only a short length of catalyst bed is required, in which case the injection-port liners of commercial chromatographs can serve as a suitable pre-column.[60] Functional groups containing O, N, S, and halogens are replaced by H atoms, and unsaturated groupings are hydrogenated, so that all eluates with the exception of six-membered cycloalkanes, which yield aromatics, yield only saturated hydrocarbon derivatives. These products are readily separated and identified, so the carbon skeleton of the original material is revealed. For example, the identification of naturally occurring steroids and sterols,[61] and of steroid drugs,[62] is greatly facilitated by their reduction to the basic hydrocarbon skeleton.

A suitable catalyst can be prepared by evaporating to dryness (in a rotary evaporator) a palladium chloride solution in 5% acetic acid in contact with a non-adsorptive solid support. It is essential to produce a neutral or alkaline catalyst, since traces of acid cause the breakdown of chains longer than C_9. Acidity is avoided by adding sodium carbonate to the catalyst solution. The final catalyst is prepared by reduction in hydrogen, and consists of about 1% by weight of palladium metal. It is operated at a temperature of several hundred degrees C.

The hydrocarbons produced from various functional groups are shown in Table 5.3. It can be seen that some groups yield the parent hydrocarbon, while others also yield the next lower homologue. Also, thermal cracking of some materials occurs, the fragment products of which are sometimes very useful in identification. Figure 5.8 shows schematically how three linear isomers of heptanone can be distinguished from the products produced by bond cleavage at either side of the carbonyl group.[67]

The technique has been used with compounds up to C_{30}. However, with such high molecular weight compounds care must be taken to eliminate 'memory' effects due to adsorption on the catalyst.

The products of reaction of most types of compound do not depend on the temperature of the catalyst. However, interconversion of six-membered cyclo-aliphatics and aromatics is temperature-dependent. Thus at 200 °C

TABLE 5.3   Products of carbon-skeleton chromatography[59] (catalyst of 1% Pd at 300 °C; flow rate 20 ml/min)

Compounds giving parent exclusively	Reaction
Paraffinic hydrocarbon	None
Unsaturated compound	Multiple bonds satd.
Halogenated compound	C—X bond cleaved
Alcohol (sec. or tert.)	C—O bond cleaved
Ester (alcohol part sec. or tert.)	C—O bond cleaved
Ether (sec. or tert.)	C—O bond cleaved
Ketone	C=O bond cleaved
Amine (sec. or tert.)	C—N bond cleaved
Amide (NH attached to sec. or tert. C)	C—N bond cleaved

Compounds giving parent and/or next lower homologue	Reaction
Aldehyde	$RCHO \rightarrow RH, RCH_3{}^a$
Acid	$RCOOH \rightarrow RH, RCH_3{}^a$
Anhydride	$(RCO)_2O \rightarrow RH, RCH_3{}^a$
Alcohol (primary)	$RCH_2OH \rightarrow RH, RCH_3{}^a$
Ester (C—O attached to primary C)	$R'COOCH_2R \rightarrow RH, RCH_3; R'H, R'CH_3{}^a$
Ether (primary)	$RCH_2OCH_2R \rightarrow RH, RCH_3$
Amide (NH attached to primary C)	$R'CONHCH_2R \rightarrow RH, RCH_3; R'H, R'CH_3{}^a$

^{a} Little or none of this parent hydrocarbon is obtained.

aromatisation predominates whilst at 360 °C hydrogenation is favoured. The technique is equally applicable to microgram amounts of sample and to micropreparative work.[63] An example of its use is to distinguish between the two dimethyl-substituted cyclohexenecarboxylic acids (A) and (B). The

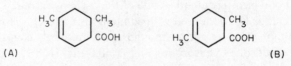

(A)   (B)

product of carbon skeleton chromatography of the unknown at 300 °C was *m*-xylene, and the acid was therefore identified as compound (A).[59]

A related method is reductive dealkylation in which alkyl side-chains are stripped from aromatic nuclei by catalytic reduction with hydrogen.[64] Interfering groups such as COOH and SO_3H must first be removed.

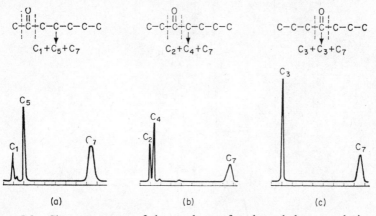

FIGURE 5.8 Chromatograms of the products of carbon skeleton analysis of (a) heptan-2-one, (b) heptan-3-one, and (c) heptan-4-one. (From ref. 67)

5.6   REACTION RATE METHODS

In principle, the rate at which a known type of compound reacts with an appropriate reagent may enable it to be identified exactly. In practice, this approach to identification in GC is hardly ever used, probably because there are many more convenient methods of identification available.

One method which has received some attention involves the reaction of an eluate as it is eluted with a reactive column substrate. Gil-Av and Herzberg-Minzley[65] coined the name 'partial subtraction chromatography' for this use of reactive stationary phases, and they pointed out that a study of relative peak sizes as a function of flow rate would give a measure of the reactivity of the component concerned toward the stationary phase. By way of illustration they presented results obtained for a variety of conjugated dienes chromatographed on a dienophilic stationary phase, chloromaleic anhydride (Figure 5.9). There is no reason why the stationary phase itself need be reactive, and greater flexibility might well ensue if conventional liquid phases were used as solvents for more interesting reagents.

An even more flexible technique would be one involving a pre-column reaction chamber containing a suitable reagent from which small samples of eluate could be withdrawn at intervals for peak-height analysis. Individual eluate peaks could, if necessary, be diverted into the chamber for reaction-rate measurement as required. In general, however, reaction-rate methods in GC are unlikely to be very useful.

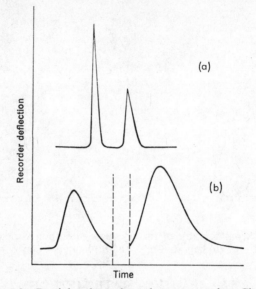

FIGURE 5.9   Partial subtraction chromatography. Chromatograms of *cis*- and *trans*-penta-1,3-diene on a stationary phase containing chloromaleic anhydride, demonstrating its faster reaction with the *trans*-isomer. (a) Fast flow (77 ml/min); (b) slow flow (11·4 ml/min). The first peak in each case is for the *trans*-isomer. (From ref. 65)

REFERENCES

1. J. T. Walsh and C. Merritt, 'Qualitative functional group analysis of gas chromatographic effluents', *Analyt. Chem.*, **32**, 1378 (1960).
2. (*a*) V. G. Berezkin, 'Reaction gas chromatography in analysis', *Russ. Chem. Rev.*, **34**, 470 (1965). (*b*) V. G. Berezkin, *Analytical Reaction Gas Chromatography*, Plenum Press, New York (1968).
3. B. Casu and L. Cavallotti, 'A simple device for qualitative functional group analysis of gas chromatographic effluents', *Analyt. Chem.*, **34**, 1514 (1962).
4. J. E. Hoff and E. D. Feit, 'New technique for functional group analysis in gas chromatography', *Analyt. Chem.*, **36**, 1002 (1964).
5. R. Bassette, S. Özeris, and C. H. Whitnah, 'Gas chromatographic analysis of head space gas of dilute aqueous solutions', *Analyt. Chem.*, **34**, 1540 (1962).
6. M. Beroza and B. A. Bierl, 'Rapid determination of olefin position in organic compounds in the microgram range by ozonolysis and gas chromatography', *Analyt. Chem.*, **39**, 1131 (1967).

7. V. L. Davison and H. J. Dutton, 'Microreactor chromatography—quantitative determination of double bond position by ozonolysis-pyrolysis', *Analyt. Chem.*, **38**, 1302 (1966).

8. E. C. Nickell and O. S. Privett, 'A simple, rapid micromethod for the determination of the structure of unsaturated fatty acids via ozonolysis', *Lipids*, **1**, 166 (1966).

9. R. R. Allen, 'Selective removal of aldehydes from complex mixtures during gas chromatography', *Analyt. Chem.*, **38**, 1287 (1966).

10. G. F. Spencer, R. Kleiman, F. R. Earle, and I. A. Wolff, 'Rapid micromethod for location of ene-yne and α-hydroxy conjugated diene systems in straight-chain compounds', *Analyt. Chem.*, **41**, 1874 (1969).

11. J. W. Ralls, 'Rapid methods for semiquantitative determination of volatile aldehydes, ketones and acids. Flash exchange gas chromatography', *Analyt. Chem.*, **32**, 332 (1960).

12. L. A. Jones and R. J. Monroe, 'Flash exchange method for quantitative gas chromatographic analysis of aliphatic carbonyls from their 2,4-dinitrophenylhydrazones', *Analyt. Chem.*, **37**, 935 (1965).

13. L. Gasco, R. Barrera, and F. de la Cruz, 'Gas chromatographic investigations of the volatile constituents of fruit aromas', *J. Chromatog. Sci.*, **7**, 228 (1969).

14. I. C. Cohen and B. B. Wheals, 'An EC–GC method for the substituted urea and carbamate herbicides as 2,4-dinitrophenylhydrazone derivatives of their amine moieties', *J. Chromatog.*, **43**, 233 (1969).

15. S. H. Langer and P. Pantages, 'Peak-shift technique in gas–liquid chromatography: trimethylsilyl ether derivatives of alcohols', *Nature*, **191**, 141 (1961).

16. M. W. Anders and G. J. Mannering, 'New peak-shift technique for gas–liquid chromatography. Preparation of derivatives on the column', *Analyt. Chem.*, **34**, 730 (1962).

17. S. J. Mulé, 'Determination of narcotic analgesics in human biological materials. Application of ultraviolet spectrophotometry, thin layer and gas chromatography', *Analyt. Chem.*, **36**, 1907 (1964).

18. P. S. Jaglan, R. B. March, and F. A. Gunther, 'Column esterification in the gas chromatography of the desalkyl metabolites of methyl parathion and methyl paraoxon', *Analyt. Chem.*, **41**, 1671 (1969).

19. A. E. Pierce, *Silylation of Organic Compounds*, Pierce Chemical Co., Rockford, Illinois, U.S.A. (1968).

20. G. R. Jamieson and E. H. Reid, 'The analysis of oils and fats by gas chromatography. VII. Separation of long-chain fatty alcohols as their trifluoroacetyl and trimethylsilyl derivatives', *J. Chromatog.*, **40**, 160 (1969).

21. K. L. Leibman and E. Ortiz, 'Thin layer and gas chromatography of trimethylsilyl ethers of glycols', *J. Chromatog.*, **32**, 757 (1968).

22. G. G. Esposito and M. H. Swann, 'Gas chromatographic determination of polyhydric alcohols in oils and alkyd resins by formation of trimethylsilyl derivatives', *Analyt. Chem.*, **41**, 1118 (1969).

23. G. E. Martin and J. S. Swinehart, 'Comparison of gas chromatography of methyl and trimethylsilyl esters of alkanoic and hydroxypolycarboxylic acids', *J. Gas Chromatog.*, **6**, 533 (1968).

24. F. A. Vandenheuvel and A. S. Court, 'Reference high-efficiency nonpolar packed columns for the gas liquid chromatography of nanogram amounts of steroids. Part 1. Retention time data', *J. Chromatog.*, **38**, 438 (1969).
25. C. C. Sweeley, R. Bentley, M. Makita, and W. W. Wells, 'Gas liquid chromatography of trimethylsilyl derivatives of sugars and related substances', *J. Amer. Chem. Soc.*, **85**, 2497 (1963).
26. W. E. Wilson, S. A. Johnson, W. H. Perkins, and J. E. Ripley, 'Gas chromatographic analysis of cardiac glycosides and related compounds', *Analyt. Chem.*, **39**, 40 (1967).
27. C. T. Bishop, 'Separation of carbohydrate derivatives by gas liquid chromatography', in *Methods of Biochemical Analysis*, Vol. 10 (ed. D. Glick), Interscience, New York (1962).
28. W. C. Ellis, 'Liquid phases and solid supports for gas liquid chromatography of trimethylsilyl derivatives of monosaccharides', *J. Chromatog.*, **41**, 334 (1969).
29. D. B. Gower and B. S. Thomas, 'Gas liquid chromatography of androst-16-enes as trimethylsilyl and chloromethyldimethylsilyl ethers', *J. Chromatog.*, **36**, 338 (1968).
30. T. Furuya and H. Kojima, 'Gas liquid chromatography of valerian sesquiterpenoids', *J. Chromatog.*, **29**, 341 (1967).
31. R. J. Argauer, 'Rapid procedure for chloroacetylation of microgram quantities of phenols and detection by electron capture gas chromatography', *Analyt. Chem.*, **40**, 122 (1968).
32. F. Drawert, R. Felgenhauer, and G. Kupfer, 'Reaction gas chromatography', *Angew. Chem.*, **72**, 555 (1960) [*Chem. Abs.*, **55**, 60a (1961)].
33. F. Drawert, 'Reaction gas chromatography', *Abhandl. Deut. Akad. Wiss. Berlin, Kl. Chem., Geol., Biol.* [*Chem. Abs.*, **60**, 4439b (1964)].
34. F. Drawert, H.-J. Kuhn, and A. Rapp, 'Reaction gas chromatography. III. Gas chromatographic determination of lower fatty acids in the stomach of leaf-eating monkeys (Colobinae)', *Z. Physiol. Chem.*, **329**, 84 (1962) [*Chem. Abs.*, **58**, 4802a (1963)].
35. D. van Wijngaarden, 'Rapid preparation of fatty acid esters from lipids for gas chromatographic analysis', *Analyt. Chem.*, **39**, 849 (1967).
36. J. J. Bailey, 'Determination of aliphatic and aromatic acids by pyrolysis of their tetraalkylammonium salts', *Analyt. Chem.*, **39**, 1485 (1967).
37. J. W. Schwarze and M. N. Gilmour, 'Gas chromatographic determination of C_1 to C_7 monocarboxylic acids and lactic acid by pyrolysis of the tetrabutylammonium salts', *Analyt. Chem.*, **41**, 1686 (1969).
38. J. W. Ralls, 'Flash exchange gas chromatography for the analysis of potential flavour components of peas', *J. Agric. Food Chem.*, **8**, 141 (1960).
39. I. R. Hunter, 'Extension of flash exchange gas chromatography to ethyl esters of higher organic acids', *J. Chromatog.*, **7**, 288 (1962).
40. J. Janak, J. Novak, and J. Sulovsky, 'Separation of substituted malonic acid esters by gas chromatography, and a new method of identification', *Coll. Czech. Chem. Comm.*, **27**, 2541 (1962) [*Chem. Abs.*, **58**, 2827b (1963)].
41. A. Kuksis, 'Newer developments in determination of bile acids and steroids by gas chromatography', in *Methods of Biochemical Analysis*, Vol. 14 (ed. D. Glick), Interscience, New York (1966), p. 383.

42. I. R. Hunter, K. P. Dimick, and J. W. Corse, 'Determination of amino acids by ninhydrin oxidation and gas chromatography', *Chem. and Ind.* (London), 294 (1956).

43. A. Zlatkis and J. F. Oró, 'Amino acid analysis by reactor gas chromatography', *Analyt. Chem.*, **30**, 1156 (1958).

44. A. Zlatkis, J. F. Oró, and A. P. Kimball, 'Direct amino acid analysis by gas chromatography', *Analyt. Chem.*, **32**, 162 (1960).

45. B. Weinstein, 'Separation and determination of amino acids and peptides by gas liquid chromatography', in *Methods of Biochemical Analysis*, Vol. 14 (ed. D. Glick), Interscience, New York (1966), p. 203.

46. S. Makisumi, C. H. Nicholls, and H. A. Saroff, 'The influence of esterifying and acetylating groups on the retention times of amino acid derivatives in gas chromatography', *J. Chromatog.*, **12**, 106 (1963).

47. W. M. Lamkin and C. W. Gehrke, 'Quantitative gas chromatography of amino acids. Preparation of *n*-butyl N-trifluoroacetyl esters', *Analyt. Chem.*, **37**, 383 (1965).

48. A. Islam and A. Darbre, 'Gas–liquid chromatography of trifluoroacetylated amino acid methyl esters', *J. Chromatog.*, **43**, 11 (1969).

49. M. D. Waterfield and A. Del Favero, 'Purification of N-trifluoroacetyl amino acid n-butyl esters for analysis by gas chromatography', *J. Chromatog.*, **40**, 294 (1969).

50. W. H. McCurdy and R. W. Reiser, 'Analysis of fatty amines as trifluoroacetyl derivatives', *Analyt. Chem.*, **38**, 795 (1966).

51. R. A. Dove, 'Separation and determination of aniline and toluidine, xylidine, ethylamine and N-methyltoluidine isomers by gas chromatography of the N-trifluoroacetyl derivatives', *Analyt. Chem.*, **39**, 1188 (1967).

52. W. J. Irvine and M. J. Saxby, 'Gas chromatography of primary and secondary amines as their trifluoroacetyl derivatives', *J. Chromatog.*, **43**, 129 (1969).

53. W. J. A. Vanden Heuvel, 'Gas liquid chromatography of the trimethylsilyl derivatives of several amines of biological interest', *J. Chromatog.*, **36**, 354 (1968).

54. W. J. A. Vanden Heuvel, W. L. Gardiner, and E. C. Horning, 'Characterization and separation of amines by gas chromatography', *Analyt. Chem.*, **36**, 1550 (1964).

55. B. Smith, R. Ohlson, and G. Larson, 'Relative retention times of C_2 to C_7 saturated and unsaturated hydrocarbons and their use for the determination of boiling point and hydrocarbon type', *Acta Chem. Scand.*, **17**, 436 (1963).

56. J. Franc and V. Koloušková, 'Structural analysis of organic substances by means of hydrogenation combined with gas chromatography', *J. Chromatog.*, **17**, 221 (1965).

57. I. R. Klesment, 'Hydrogenation and dehydrogenation of gas chromatographically separated substances in a two-component carrier gas determination of the concentration of the reacting gas', *J. Chromatog.*, **31**, 28 (1967).

58. A. B. Littlewood and W. A. Wiseman, 'A servomechanism detector for the quantitative measurement of gas-chromatographic zones of hydrogenatable material', *J. Gas Chromatog.*, **5**, 334 (1967).

59. M. Beroza and R. Coad, 'Reaction gas chromatography', in *The Practice of Gas Chromatography* (ed. L. S. Ettre and A. Zlatkis), Interscience, New York (1967), p. 461; *J. Gas Chromatog.*, **4**, 199 (1966).
60. B. A. Bierl, M. Beroza, and W. T. Ashton, 'Reaction loops for reaction gas chromatography. Subtraction of alcohols, aldehydes, ketones, epoxides, and acids and carbon-skeleton chromatography of polar compounds', *Mikrochim. Acta*, 637 (1969).
61. P. M. Adhikary and R. A. Harkness, 'Determination of carbon skeleton of microgram amounts of steroids and sterols by gas chromatography after their high temperature catalytic reduction', *Analyt. Chem.*, **41**, 470 (1969).
62. P. M. Adhikary and R. A. Harkness, 'Production of the parent hydrocarbons from steroid drugs and their separation by gas chromatography', *J. Chromatog.*, **42**, 29 (1969).
63. R. G. Brownlee and R. M. Silverstein, 'A micro-preparative gas chromatograph and a modified carbon skeleton determinator', *Analyt. Chem.*, **40**, 2077 (1968).
64. J. Franc, J. Senkyova, F. Mikes, and K. Placek, 'Identification of an alkyl group and its position in aromatic substances by reaction gas chromatography', *J. Chromatog.*, **43**, 1 (1969).
65. E. Gil-Av and Y. Herzberg-Minzley, 'Partial subtraction chromatography', *J. Chromatog.*, **13**, 1 (1964).
66. K. B. Eik-Nes and E. C. Horning, *Gas Phase Chromatography of Steroids*, Springer-Verlag, Berlin (1968).
67. M. Beroza and R. Sarmiento, 'Determination of the carbon skeleton and other structural features of organic compounds by gas chromatography', *Analyt. Chem.*, **35**, 1353 (1963).
68. J. Guillot, H. Boltazzi, A. Guyot, and Y. Trambouzé, 'A gas chromatographic selective detector for hydrogenatable material', *J. Gas Chromatog.*, **6**, 605 (1968).

CHAPTER SIX

PYROLYSIS GAS CHROMATOGRAPHY AND OTHER DEGRADATIVE METHODS OF IDENTIFICATION

In pyrolysis gas chromatography (PGC) a sample for analysis is pyrolysed, and the resulting reaction products are separated by GC in a concerted operation. There are two main reasons for using the technique: (a) to study mechanisms of pyrolysis, and (b) to provide a fingerprint chromatogram (a pyrogram) for identification of the sample. In this chapter we shall be largely concerned with the latter aspect, although it is clear that a knowledge of pyrolytic mechanisms forms a useful background for PGC. At present, however, PGC is most powerful as an empirical technique of identification, and in this mode it is analogous to mass spectroscopy (MS), in which, of course, decomposition is due to electron impact.

There has been a rapid growth of interest in PGC in the last few years owing principally to its usefulness in the identification of solids such as polymers and plastics. These intractable materials, often insoluble and inert, frequently exhibit characteristic pyrograms; identification then follows either from a purely empirical comparison of the pyrogram with those of standard solids, or by consideration of the peaks identified in the pyrogram in the light of known mechanisms of pyrolysis.

An important application of PGC which has been relatively neglected is its use in the identification of volatile materials and, in particular, of GC peaks. It is this aspect which is most relevant to the theme of this book, although for completeness the identification of solids is briefly considered.

6.1   GENERAL CONSIDERATIONS

The primary requirement of any PGC technique is that it should yield the same pyrograms for repeat experiments, and this is entirely a feature of the apparatus.

Demands on the gas chromatograph are not very exacting. A column which enables the pyrogram to be obtained quickly is useful, particularly when using PGC on individual peaks eluted from another column. High chromatographic resolution is rarely required, since in most applications of PGC to identification it is simply the overall pattern of the pyrogram which is of importance. Furthermore, resolution is often limited by the large width of the injected peak containing pyrolysis products.

The characteristics of the pyrolysis unit are of fundamental importance in any PGC set-up. A critical factor is the temperature–time profile of the unit, and hence the sample, since pyrolytic mechanisms depend qualitatively and quantitatively on temperature. Such profiles have been obtained oscilloscopically by monitoring filament emission with a photomultiplier or by using a low mass thermocouple.[1,2] The behaviour of the sample in the hot region—how much of it remains in the vicinity, and for how long—is also of importance, as are long-term changes (ageing) of the pyrolysis unit. Ageing effects are usually of greatest significance in surface-catalysed processes, and it is therefore desirable to minimise such reactions. Ideally, the pyrogram should not depend on size of sample, or extent of decomposition, and it should be uniquely characteristic of the sample compound.

In general, all these factors cannot be simultaneously satisfied, and compromise must be sought. The most that can be hoped for in fingerprint PGC is good reproducibility of pyrolysis conditions, which, for approximately similar samples as regards size, will lead to similar pyrograms. In general, pyrograms should be used for identification purposes only when obtained in a well characterised apparatus using a standardised procedure. At the present time, the most serious limitation of PGC is that results from different laboratories cannot be compared with any certainty.

6.2. PYROLYSIS UNITS

The many varied types of pyrolysis unit reported up to late 1965 have been comprehensively reviewed.[3] The number of different designs of pyrolysis unit does not fall far short of the number of publications on the subject. However, most designs fall into one of two categories, namely, filament or microreactor types.

6.2a Pulsed filament pyrolysis units

Resistive heating

These units consist of a wire, a ribbon, or a foil which is very rapidly raised to a high temperature by resistive heating. A timer is set to allow the pyrolysis to proceed for a predetermined time, the products of the reaction being swept into the chromatograph. The technique, sometimes called 'flash pyrolysis', is most useful for involatile samples, which are usually contained in a suitable cavity or dish fashioned from the filament. Alternatively, the sample is placed in a small boat or tube, or wrapped in glass-fibre paper, and inserted in a coiled filament. In another method, the sample is applied to the filament in solution and the solvent is evaporated to leave a thin film of sample. This technique is especially suited to microgram quantities of sample. Reproducibility is high with *very thin* films, and therefore, in order to allow the use of reasonably large sizes of sample, filaments with large surface areas are needed. Spirals of foil have been used,[4] and also ribbon filaments; the latter have the added advantage of a more even distribution of temperature. The pyrograms of thin (200 Å) polymer films confined between two engraved lines on a ribbon have been found to be highly reproducible.[5]

The filament type of pyrolysis unit has a number of disadvantages. There is inevitably a lag of a few seconds before the filament and its sample reach an equilibrium temperature, and during this interval pyrolysis proceeds over a range of temperature. In many respects this does not matter as long as the warm-up profile is the same for all samples. However, this condition is difficult to achieve over a range of differing samples, particularly when they are large (~ 1 mg), since the heat-transfer parameters will vary from sample to sample. In the thin-film technique the thermal capacity of the sample is much smaller than that of the filament, and such problems are not therefore so acute.

Warm-up times can be reduced, and hence the reproducibility of the temperature–time profile improved, by applying a high initial voltage to the filament and causing it to fall as the filament temperature approaches a predetermined value.[6] The equilibrium temperature itself is a function of the sample as well as the pyrolysis unit, and it is desirable to have some independent means of checking its value. Thermocouples and optical pyrometers are among the usual methods employed.

The shape of the filament and the geometry of the chamber which contains it may also be critical in determining pyrolysis conditions. On firing the filament, a pressure surge is created in the chamber, and the products of pyrolysis escape from the sample in all directions. Products which escape upstream may come into contact with the filament on their downstream flow, and be subjected to further decomposition. Such effects are less likely in a large chamber but chromatographic resolution may then be seriously impaired.

Unfortunately, even when a pyrolysis unit satisfactory in terms of thermal characteristics has been obtained, its long-term performance is not assured; constant current control is not sufficient to achieve a constant, reproducible temperature rise and equilibrium temperature, since these are affected by such factors as ageing and carbonisation of the filament. It is therefore necessary to perform regular checks of performance, such as carrying out temperature checks and standardised pyrolytic experiments.

Another difficulty which may be encountered is catalysis by the filament. This is not a problem by itself but is likely to affect overall, long-term reproducibility, since significant changes in the surface of the filament are quite likely. Further, it should be noted that in thin-film techniques the structure of the film may be different from that of the bulk sample, and care should therefore be taken to standardise coating procedure.

Inductive heating

The most promising technique so far devised for the PGC of solids is undoubtedly that which uses inductive heating. The solid is applied as a thin film to a ferromagnetic wire conductor (typically 20×0.5 mm) which is rapidly brought up to, but not beyond, its Curie temperature by radio-frequency (typically 450 kHz) heating via a copper coil. The reported repeatability is good.[2]

Figure 6.1(a) shows an induction pyrolysis unit, while Figure 6.1(b) gives reported temperature–time curves for a variety of ferromagnetic materials.[7] It can be seen that a wide choice of final temperatures is available, and that typical warm-up times are as low as 0.03 second. Under these conditions the extent of pyrolysis which takes place below the final temperature can be very small. However, if high final temperatures are used, substantial sample decomposition can occur in a few milliseconds. The power output of the induction generator is an important factor in determining the warm-up time. A commercial induction pyrolyser employing less than 50 W has been reported to have a warm-up time of as much as about 1.5 seconds, whereas a 2.5 kW generator had a warm-up time of 0.12 second.[1,2]

As described above, this technique is applicable to microgram quantities of sample. It can be satisfactorily extended to larger quantities (~ 1 mg) by using a mixture of sample and finely divided iron in place of the ferro-magnetic conductor.[8] Clearly, such a device involves the same thermal problems as those encountered with other large-sample methods.

6.2b   Other pulsed pyrolysis units

In addition to the pulsed pyrolysis units described above, a number of others have been devised, including those using dielectric breakdown of the sample[9] and pyrolysis within a carbon arc.[10] These particular methods are not very useful, since they degrade all materials into small uncharacteristic molecules. Of much greater promise, but as yet not extensively developed, are those units which rely upon high-intensity thermal radiation from a carbon arc[11] or xenon flash tube.[12] In this context, the possibility of using laser radiation to achieve fast, localised heating is perhaps worth exploring. (For recent papers on this subject, see refs. 39 and 40.)

6.2c   Microreactors

Many of the problems of the filament pyrolysis unit are due to its thermal transient behaviour. The microreactor, consisting essentially of a tiny flow-through furnace, has the advantage of presenting a stable thermal

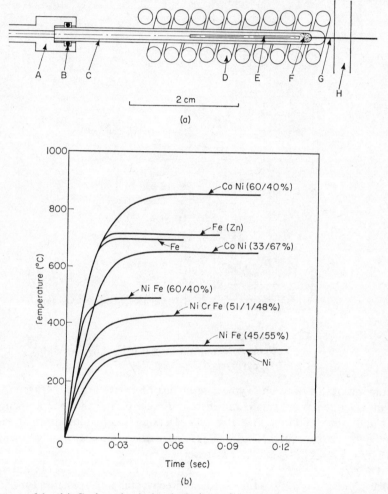

FIGURE 6.1   (a) Curie point inductively heated pyrolysis unit. (A) Carrier gas connector; (B) O-ring seal; (C) glass capillary; (D) copper induction coil operated at 450 kHz; (E) ferromagnetic conductor 20 mm long and 0·5 mm outside diameter; (F) sintered glass beads; (G) Pt–Ir capillary, 0·3 mm inside diameter; (H) injection port. (From ref. 7.) (b) Temperature–time curves for various ferromagnetic conductors in apparatus (a). (From ref. 7)

environment into which the sample is introduced. It is suitable for use with solids, liquids, gases, and vapours.

Figure 6.2 shows a highly developed microreactor[13,14] which consists of a long thin tube ($10^3 \times 1$ mm) wound helically round a silver core, and surrounded by a silver jacket to provide high thermal inertia and even

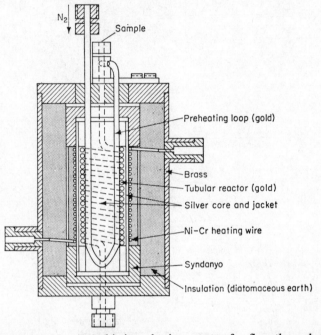

FIGURE 6.2   A sophisticated microreactor for flow-through pyrolysis. (From ref. 13)

temperature distribution. Typical residence times are around 20 seconds. Wall-layer effects are small, and therefore every part of the sample spends virtually the same time in the reactor. Heating and cooling times at exit and entry are short compared with residence times, while thermal gradients are minimised by preheating the carrier gas.

The large surface/volume ratio of the tube is a potential source of catalytic effects, particularly if the tube is carbonised. Apparently, reaction tubes of copper, silver, or gold give similar results, while pyrograms obtained with platinum or stainless steel tubes do not. Tubes of precious metal have the advantage that any carbon formed during pyrolysis can easily be burnt off. Clearly, the extent of carbon formation will depend on the type of sample being pyrolysed, and also upon the extent of decomposition involved.

A variety of methods of sample introduction are available. Volatile materials are merely injected into the carrier gas stream ahead of the microreactor. Liquids should be vaporised before entering the micro-reactor, by using e.g. a heated injection port. Solids may be introduced in small boats which are manipulated into the microreactor by devices such as a push-rod operated magnetically from outside.[15] Alternatively, solids may be simply dropped into the microreactor; involatile liquids can be similarly introduced either as droplets or absorbed into tiny balls of quartz wool.

One of the disadvantages of microreactors is that the introduction of the sample produces a fall in temperature. A large thermal inertia together with good thermal transfer is required to minimise this disturbance. In unfavour-able cases, when samples take a long time to reach the temperature of the microreactor, PGC peaks may be broad and poorly resolved. This effect is made worse by the fact that reactor volumes are usually rather large. Packed reactors are sometimes used to improve thermal contact, although it should be noted that the heterogeneous processes which inevitably occur in any microreactor may be accentuated in such units. Another feature of many poorly designed flow-through reactors is that there is a thermal gradient along their length. The sample and its pyrolysis products there-fore decompose over a range of temperature. The reactor of Figure 6.2 has a very small temperature gradient.

6.3   PYROLYSIS GAS CHROMATOGRAPHY AS A MEANS OF IDENTIFICATION

6.3a   Pyrolysis gas chromatography of solids

Although the PGC of solids has been extensively developed in recent years, good inter-laboratory reproducibility has not been attained, particularly when results obtained with filament and microreactor techniques are com-pared. However, despite the many shortcomings of the experimental techniques, much valuable information has come from these studies. This is particularly true of the 'fingerprint' approach with a standard apparatus, which is invaluable for the identification of high polymers, copolymers, and polymer mixes.[2,16] Mechanistic studies of the pyrolysis of polymers and of complex organic molecules are on a less sure footing. Unless scrupulous precautions are taken, the products of pyrolysis can present a very mis-leading picture of the molecular structure of the sample.

Inductively heated filament units, or resistively heated ribbons, with very thin films give the best results in the PGC of solids. Small samples, down to about 10^{-8} g, are necessary, and this requires the use of detectors of high sensitivity.

PGC appears to have a promising future as a means of characterising biological materials such as proteins and microorganisms.[17-22] Characterisation of microorganisms can be achieved only by paying great attention to the purity of the sample, since differences in the pyrograms of such organisms are mainly quantitative rather than qualitative.[19] Nonetheless, the future of chemotaxonomy based on PGC seems good, and it is likely to become an important biological technique.

It is possible that PGC coupled with mass spectrometry will be used on the first Mars landing in order to test for the presence of biological materials. Consideration of some characteristic pyrolysis fragments of biological molecules indicates that PGC can distinguish between living biological material and organic matter such as that found in meteorites or fossil shales.[23]

6.3b Pyrolysis gas chromatography of volatile samples

PGC has achieved prominence for the identification of polymeric solids because it is one of the few techniques suitable for such work. In contrast, comparatively little attention has been given to the PGC of volatile substances, for which many alternative methods of identification are available. It should be noted, however, that the use of GC to analyse reaction products has revolutionised kinetic studies of gaseous chemical reactions, including pyrolyses. Nevertheless, in these experiments the pyrolytic and analytical functions are separate, and the direct transfer by carrier gas of pyrolysis products from a microreactor to a gas chromatographic column has only rarely been used. Such work may well receive more attention in the future as the need for simple techniques of identification of GC peaks becomes more widely felt, for which the combination of GC–pyrolysis–GC, discussed further in the next section, offers attractive possibilities.

Sample handling and pyrolysis unit requirements for the PGC of volatile substances differ significantly from those for solid samples. The major difference is the great simplification of the heat-transfer/heating-rate problem. Further, the undecomposed portion of the sample can itself be chromatographed, which is not only useful in indicating the extent of pyrolysis, but also means that the whole sample is removed from the reactor, thereby improving reproducibility.

Originally pioneered with a filament unit,[24] and developed later with a microreactor, PGC enables the yields of such products as H_2, CO, CO_2, and CH_4 to be related in a semiquantitative way to the functional groups present in a variety of aromatic compounds. For example, it is found that phenols give a substantial yield of CH_4, while nitro groups give none. Clearly, the microreactor type of pyrolysis unit is preferable for the PGC of volatile substances, and is most commonly used at present. The technique has successfully been applied to the detection of aromatic rings,[25] the identification of C_5 molecules in food odours[26] (see Figure 6.3), and the

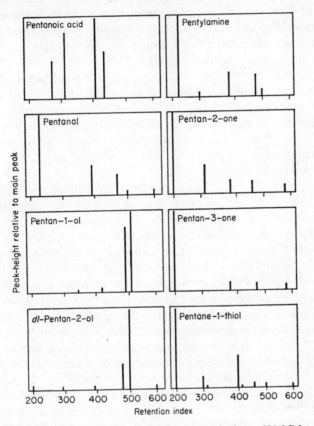

FIGURE 6.3 Pyrograms obtained from pyrolysis at 600 °C in a quartz tube packed with firebrick of a variety of monofunctional C_5 straight-chain compounds. Relative peak-heights are plotted against Kovats retention indices on Apiezon-L at an unspecified temperature. (From ref. 26)

identification of alcohols.[27] An important aspect of this work has been the PGC of gas chromatographic peaks, which is discussed in more detail below (see pp. 123—130).

Experimentally, the most obvious way of combining a microreactor and GC is by direct linkage. There are, however, a number of other ways which

E

have some advantages. In one such technique, which is perhaps of particular use in process control, since it is very wasteful of sample material, the sample is leaked continuously into the microreactor, and samples of the reaction products are injected at intervals into the chromatograph by means of a conventional sampling valve. Hence, carrier flow rates through the microreactor and gas chromatograph can be independently controlled. This arrangement enables the residence time of the sample in the microreactor to be varied without altering the GC conditions, and hence, if required, rate constants for the production of various peaks of the pyrogram can be obtained.[13] Although such rate data may be useful for identification purposes, they are of limited kinetic significance since pyrolyses usually involve many secondary reactions. Furthermore, the rate depends on the reactant concentration unless the pyrolysis happens to have a kinetic order of 1·0, which is very uncommon, despite a widespread impression to the contrary.

The addition of a subsidiary carrier gas supply after the pyrolyser permits some independence of flow rates for pyrolysis and GC, even with a simple directly coupled system. Another way of separating the pyrolysis and GC functions in PGC is to trap the products of reaction (see Chapter Nine), and then to inject them into the chromatograph at leisure in the normal way.

The pyrogram obtained at any temperature is determined by a large number of kinetic factors, of which two are particularly important. First, the extent of decomposition; at low extents of reaction the product distribution tends to be more characteristic, while at high extents many products either polymerise or carbonise, or degrade to smaller, less characteristic molecules. It is therefore important to compare pyrograms of the standard and the unknown at similar extents of reaction. The second factor which has sometimes been shown to be important,[28] and is likely to have some effect in all situations, is the catalytic effect of traces of air (oxygen). It should be noted that the effect of traces of oxygen is often to accelerate or retard the pyrolytic process proper, possibly with some changes of product distribution, rather than to bring about oxidation. These and other factors demonstrate the need for close control of pyrolytic conditions, and the desirability of directly comparing the unknown with standard substances on the same apparatus.

A standard procedure using retention indices for the characterisation of hydrocarbons by PGC has been described,[29] using a pyrolysis unit similar to that of Figure 6.2, operated at 600 °C. Residence times were about 2 seconds, and extents of decomposition between 2 and 10%. A silicone oil on firebrick column was used, and calibrated for Kovats index measurements (Section 3.4) with n-alkanes. The areas of the peaks of the pyrogram were obtained by digital integration, and normalised on a molar basis with respect to the most abundant product (i.e. excluding the parent).

Pyrograms obtained with each of two distinct sets of GC conditions, with separate pyrolysis units and chromatographs, were found to be essentially identical when presented in terms of the relative number of moles in each interval of 25 Kovats units.[29] However, it should be noted that the molar-area normalisation procedure is not practicable for an unknown compound, because the identity of each peak is not then known. It therefore seems to be better to work simply with normalised areas, even though this means that the results are only applicable to the kind of detector in use. Representative results are shown in Table 6.1 for several of the hydrocarbons investigated. The dissimilarities between different compounds are striking, and mean that there should be little difficulty in distinguishing different substances.

A vivid illustration of the utility of PGC for identification is shown in Figure 6.4, where the mass spectra of 3-methylpent-2-ene and 2-ethylbut-1-ene are shown alongside the corresponding pyrograms plotted as relative peak-heights.[29] Although it would be virtually impossible to

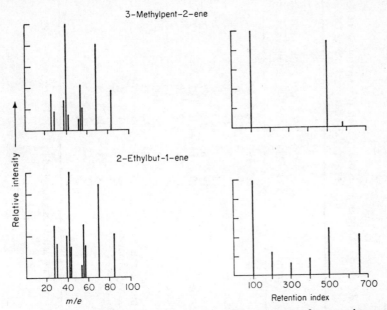

FIGURE 6.4 Comparison of mass spectra and pyrograms for two isomeric hexenes. Pyrograms were obtained by pyrolysis at 600 °C in a unit similar to that shown in Figure 6.2 coupled to a silicone oil DC 200 column at 120 °C. (From ref. 29)

TABLE 6.1 Relative FID peak areas of products from pyrolysis at 600 °C of various C₆ hydrocarbons;[a] P indicates position of parent peak; stationary phase DC 200 silicone oil, at 170 °C

Retention index interval

Compound	50/149	150/249	250/349	350/400	401/449	450/474	475/499	500/524	525/549	550/574	575/599	600/624	625/649	650/674	675/699	>700
n-Hexane	26	100	76	52												
2-Me-pentane	17	47	100	68				12				P				
3-Me-pentane	35	79	39	100				94		P						
Hex-1-ene	12	82	100	98			34		7		P					
cis-Hex-2-ene	19	65	14	97		12		100		13		P				
trans-Hex-2-ene	19	61	14	92		12		100		11		P				
cis-Hex-3-ene	17	8	2	10				100		4	P			6		
trans-Hex-3-ene	16	8	2	10				100		4	P			7		
Cyclohexene		96	26	100		2			22		24			5		5
3-Me-pent-1-ene	12	37	14	45		2	2	100				45		33	P	46
3-Me-trans-pent-2-ene	20							100		7		P				
trans,trans-Hexa-2,4-diene	2	2	1	1				5				100		P		1
cis,cis-Hexa-2,4-diene	3	2	1	2				7				100		P		2
Hex-1-yne	5	35	100	53				15				P	45	10	14	8
Hex-2-yne	22	62	7	6			4	18					3	P		
Hex-3-yne	100	7	52		14		28		7	100		P				21

[a] Calculated from raw data kindly supplied by D. L. Fanter (see ref. 29).

distinguish these two isomers by MS, there would clearly be no difficulty using PGC. In fact, the pyrograms of all 83 hydrocarbons studied by this technique were uniquely characterised,[29] with the exception of those for *cis–trans* isomers. This failing is, of course, shared by MS, but not by simple retention techniques of identification.

It is interesting to note that the observed product distributions from pyrolysis of paraffins are commonly in tolerable agreement with those expected on the basis of simple Rice–Herzfeld free radical mechanisms,[28,30] including the radical isomerisation rules proposed many years ago by Kossiakoff and Rice.[31] These rules only apply to paraffins; the mechanisms of the pyrolysis of other materials are less well understood.

Table 6.2 shows the major pyrolysis products of a number of compounds containing oxygen, together with the useful temperature range for PGC at residence times of 15 seconds in a quartz tube. The lower temperature quoted is that at which about 1% decomposition occurs, while the higher temperature corresponds to about 99% decomposition.[32]

6.3c Pyrolysis gas chromatography of gas chromatographic peaks

Most of the work reported to date on PGC of volatile compounds has been carried out by injecting samples into a pyrolysis unit coupled to a gas chromatograph. Recently, however, there has been increasing interest in systems with some means of directing the effluent from a gas chromatograph directly into the pyrolysis unit. The attractions of such a technique are obvious, and it will surely become a standard method of identification, especially for small amounts of eluate.

One version of the technique (termed here GCPGC) involves trapping the eluate from one column in a U-tube cold-trap, and subsequently sweeping it into the pyrolyser unit at the head of a second column (cf. p. 119).[27]

The need for trapping can be eliminated by directly coupling two chromatographs through a pyrolysis unit with a system of multi-port two-way taps, as shown schematically in Figure 6.5.[33] Peaks are eluted through the non-destructive detector D_1 by carrier gas supply A. As any particular peak begins to give a signal at D_1 the four-port two-way tap T_1 is turned so that the peak is diverted into the delay coil X. When the peak has been eluted, T_1 is turned back so that normal development continues in column 1, while carrier gas supply B takes over movement of the diverted peak into the helical quartz pyrolysis tube P. (The delay coil X is required so that flow conditions in the pyrolysis unit are always controlled by supply B.) From P the eluted material together with the products of pyrolysis pass through one channel of a katharometer, D_2, into a second

TABLE 6.2   Temperature ranges and products of pyrolysis for several substances[32]

Substance pyrolysed	Pyrolysis temp. range (°C)	Major pyrolysis products
$CH_3COOC(CH_3)_3$	280—340	$CH_2=C(CH_3)_2$ CH_3COOH
$(CH_2=CHCH_2)_2O$	340—440	$CH_3CH=CH_2$ $CH_2=CHCHO$
$CH_3COOCH(CH_3)CH_2CH_3$	360—460	$CH_3CH_2CH=CH_2$ $CH_3CH=CHCH_3$ (cis and trans) CH_3COOH
$CH_2=CHCH_2OCH_2CH_3$	380—480	$CH_3CH=CH_2$ CH_3CHO
$HCOOCH_2CH_2CH_2CH_3$	400—500	$CH_3CH_2CH=CH_2$ $HCOOH$
$CH_3COOCH_2CH_2CH_3$	420—500	$CH_3CH=CH_2$ CH_3COOH
$CH_3COOCH_2CH_3$	420—520	$CH_2=CH_2$ CH_3COOH
$HCOOCH_2CH_2CH_3$	420—520	$CH_3CH=CH_2$ $HCOOH$
$CH_3COOCH(CH_3)_2$	430—530	$CH_2=C(CH_3)_2$ CH_3COOH
$HCOOCH_2CH=CH_2$	460—580	$CH_3CH=CH_2$ CO_2
$CH_3CH_2COOCH_2CH=CH_2$	480—580	$CH_3CH=CH_2$ CH_3CH_2COOH CO_2
$(CH_3)_2CHOCH(CH_3)_2$	540—640	$CH_3CH=CH_2$ $CH_3CH(OH)CH_3$ CH_3COCH_3
$CH_3COCH_2CH_3$	540—640	CH_3CH_3 $CH_2=CH_2$ CO
$CH_3CH(OH)CH_3$	440—640	$CH_3CH=CH_2$ CH_3COCH_3
$CH_3CH_2OCH_2CH_3$	520—660	$CH_2=CH_2$ CH_3CH_2OH CH_3CHO
$CH_3CH_2CH_2OCH_2CH_2CH_3$	560—660	$CH_3CH=CH_2$ CH_3CH_2CHO
CH_3CH_2CHO	560—660	CH_3CH_3 $CH_2=CH_2$ CO

delay coil, Y, via a six-port two-way tap, T_2. When this is set as shown in Figure 6.5, material continues through the second channel of D_2, so that the recorder trace shows two peaks in opposite senses. From the time between these two peaks obtained in a setting-up run, it is possible to judge the best time to turn tap T_2 so that the pyrolysis mixture is back-flushed

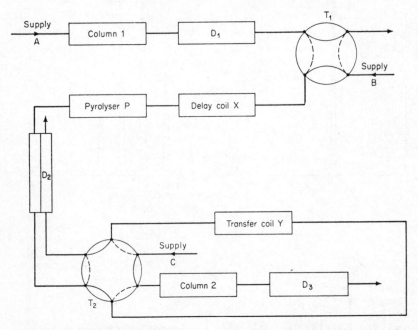

FIGURE 6.5   Schematic diagram of a unified GC–pyrolysis–GC unit with independent flow control for each chromatograph and for the pyrolysis unit. A, B, and C are carrier gas supplies; D_1, D_2, and D_3 are detectors; T_1 and T_2 are two-way multi-port taps; X and Y are helical coils 25 feet long and $\frac{3}{16}$ inch outside diameter. (From ref. 33)

from Y into the analytical column 2 by carrier gas supply C. If this is done efficiently, there should be no second peak from D_2. The pyrogram for the diverted peak is obtained from detector D_3.

This rather complicated pyrolysis transfer system has a great attraction in that the flow rate through the pyrolysis unit is independent of the flow rates for the two chromatographic columns, so that all flows can be optimised independently. The apparatus has been used in a study of the effect of pyrolysis conditions and sample size on the PGC fingerprints of

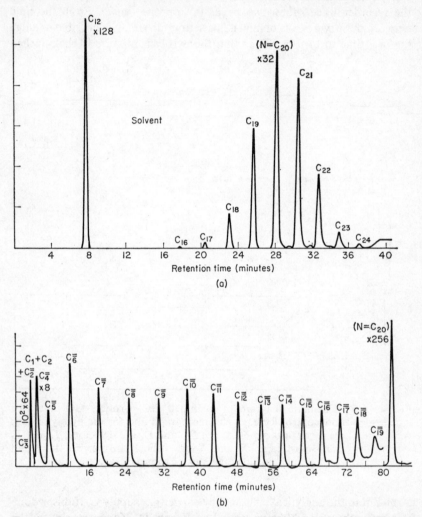

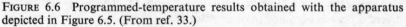

FIGURE 6.6   Programmed-temperature results obtained with the apparatus depicted in Figure 6.5. (From ref. 33.)

(a) Chromatogram of 4 μl of paraffin wax obtained with detector D_1.

(b) Pyrogram of peak 'N' of the above chromatogram obtained with detector D_3.

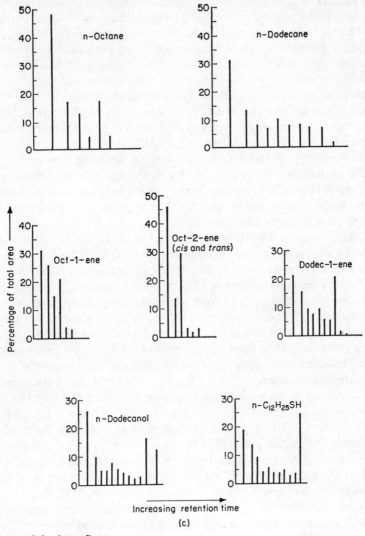

FIGURE 6.6 (contd)

(c) Normalised schematic pyrograms for a variety of C_8 and C_{12} peaks.

eluted peaks of n-alkanes, α-olefins, alcohols, mercaptans, and saturated and unsaturated methyl esters.[33] Although product distributions changed significantly with temperature and flow rate, it was found that for n-hexadecane under standard conditions they were independent of sample size. This result may not necessarily apply to all other materials. Figure 6.6 shows some programmed-temperature chromatograms and typical cracking patterns presented as normalised peak areas which were obtained using a flame ionisation detector as D_3. The analogy with MS is obvious, but at the present time cracking patterns in PGC seem to be much more variable between laboratories than is the case with MS.

The sensitivity possible with this technique is somewhat restricted by the need to use non-destructive detectors for D_1 and D_2. However, under stable standardised conditions, a trace from D_2 is not necessary, and a sensitive destructive detector could be used for D_1 in a preliminary run to establish the elution times of the various peaks.

A more serious drawback of this technique is the fact that peaks following closely behind the diverted peak cannot easily be studied in the same run. This is a common feature of all post-column identification techniques which take more than a few seconds. In general, two approaches to solving the problem can be used: either trapping and storage, or interrupted elution. Thus, in principle, it would be possible to replace T_1 (Figure 6.5) by a multi-way tap connected to a number of delay coils which could be used to store the GC peaks from column 1 between PGC runs. The alternative approach of interrupted elution is employed in the apparatus shown schematically in Figure 6.7.[34] Taps A and B are normally closed. Carrier gas flows through controller 1 and the two-way tap T as shown by the full lines. When a peak is observed at detector D_1 the four-port two-way tap T is turned so that this peak is swept into the pyrolysis unit while the carrier gas supply is cut off from column 1. Chromatographic development in column 1 is stopped by relieving the pressure at the head of the column through the pre-column by opening tap A to vent, although there is some doubt as to the need to do this.[35] (The pre-column is necessary because of slight back-development during this process; see p. 248.) After leaving the pyrolysis unit the products of pyrolysis are eluted on column 2. When the pyrogram of the selected peak is complete, column 1 is re-pressurised by closing tap A and opening tap B. After tap B is again closed the two-way tap T is rotated to continue chromatography in column 1.

Figure 6.8 shows the separation chromatogram (column 1) and the pyrogram (column 2) for a mixture of n-nonane and methyl hexanoate. It can be seen that there is no significant broadening of the second GC peak

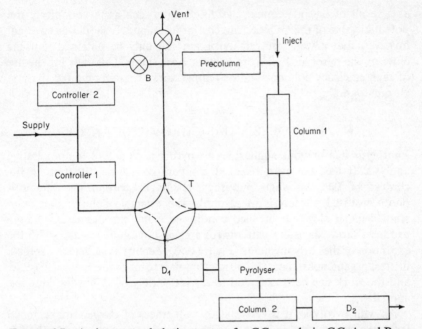

FIGURE 6.7 An interrupted-elution system for GC–pyrolysis–GC. A and B are taps, normally closed; T is a two-way four-port valve; detector D_1 is for the separation chromatogram, and D_2 for the pyrogram. (From ref. 34)

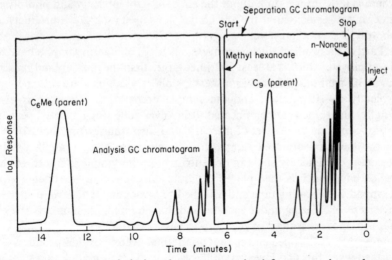

FIGURE 6.8 Interrupted-elution chromatogram (top) for two peaks, and corresponding pyrograms, obtained with the apparatus depicted in Figure 6.7. (From ref. 34)

as a result of the interrupted elution. Results such as these indicate the potential value of the GCPGC fingerprint technique. It should be stressed, however, that a well resolved pyrogram can only be obtained if peaks entering the pyrolysis unit from column 1 are sharp. This calls for the use of high efficiency columns and/or temperature programming for the first chromatograph.

6.4 PHOTOLYSIS GAS CHROMATOGRAPHY

Photolysis has obvious similarities to pyrolysis as a tool for qualitative analysis. It has, however, received comparatively little attention in the context of GC, although published results are promising. Mercury-photosensitised photolysis of alcohols and esters in solution has been studied under closely controlled conditions,[36] and chromatograms of the products show consistent patterns for a given homologous series, with the exception of the early members (Figure 6.9). Mercury is sufficiently soluble in most organic substances (1—50 micromolar) to make the method feasible, and the work has been extended to a wide variety of aldehydes, ketones, and ethers.[37]

As with pyrolysis, it is essential that all traces of oxygen are removed from the system, but it is claimed that in other respects the method is superior to pyrolysis. In particular, standardisation of the 253·6 mμ mercury resonance lamp and other photolysis conditions should be readily achieved. There is little information about the effect of light intensity and photolysis time, but it seems that at the low conversions used ($< 2\%$) the products of photolysis do not interfere.

Partly as a result of the low conversions used, milligram samples have to be employed, which is at least 10^3 times larger than the present limiting size for PGC methods. Moreover, necessary photolysis times are rather long, typically 3—10 minutes, and the sample preparation technique is also lengthy and somewhat complicated. It involves collecting an eluted peak in a trap containing a droplet of mercury, and then transferring about 10^{-3} g of the mercury-saturated sample to a 1-mm bore quartz capillary tube sealed at one end. By shaking or centrifuging, the sample is forced to the sealed end, which is then placed in liquid nitrogen while the open end is wrapped with heating tape and attached to a vacuum line. Several successive steps of freezing under vacuum and thawing under helium are carried out to remove dissolved oxygen and to ensure that all traces of sample are removed from the open end so that no thermal decomposition products are formed when this end is sealed under vacuum. This procedure takes about 5—10 minutes.

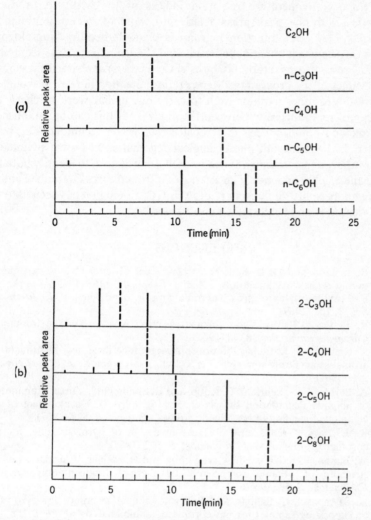

FIGURE 6.9 Schematic programmed-temperature chromatograms of products of mercury-photosensitised decomposition of (a) primary alcohols and (b) methyl secondary alcohols. Initial temperature 60 °C, rising by 5·7 °C/min. Separation achieved on a Carbowax 20M column with flow rate 42·2 ml/min. Irradiation time: (a) 10 minutes; (b) 3 minutes. Parent peaks are shown as broken lines. (From ref. 36)

The results of this photolysis GC work have been published in terms of programmed-temperature retention indices of the observed products, together with the peak areas relative to those of the most abundant product.[37] A given functional group is characterised by 'homologous' peaks at a constant retention index interval, ΔI, from the parent. 'Common' peaks are also formed (notably by esters and ethers) at characteristic absolute values of the programmed-temperature retention indices. It would be interesting if other workers were to carry out similar work to check the degree of reproducibility between laboratories. If this can be established, the method could conceivably replace established techniques for some materials. In particular, photochemical degradation has some advantages over infrared spectroscopy. It can be used, for example, to identify both the acid and alcohol moieties of an ester. Apart from the work described above, there has been little interest in photolysis GC, although it has been used for the identification of chlorinated insecticides.[38]

REFERENCES

1. R. L. Levy and D. L. Fanter, 'Measurement of temperature rise times in pyrolysis gas chromatography', *Analyt. Chem.*, **41**, 1465 (1969).
2. F. W. Willmott, 'Pyrolysis-gas chromatography of polyolefins', *J. Chromatog. Sci.*, **7**, 101 (1969).
3. R. L. Levy, 'Pyrolysis gas chromatography. A review of the technique', *Chromatographic Rev.*, **8**, 48 (1966).
4. J. Franc and J. Blaha, 'Gruppenmässige Identifizierung nichtflüchtiger aromatischer Stoffe mit Hilfe der Gaschromatographie', *J. Chromatog.*, **6**, 396 (1961).
5. A. Barlow, R. S. Lehrle, J. C. Robb, and D. Sunderland, 'Direct examination of polymer degradation by gas chromatography. II. Development of the technique for quantitative kinetic studies', *Polymer*, **8**, 523 (1967).
6. R. L. Levy, 'Trends and advances in design of pyrolysis units for gas chromatography', *J. Gas Chromatog.*, **5**, 107 (1967).
7. W. Simon, P. Kriemler, J. A. Voellmin, and H. Steiner, 'Elucidation of the structure of organic compounds by thermal fragmentation', *J. Gas Chromatog.*, **5**, 53 (1967).
8. H. Szymanski, C. Salinas, and P. Kwitowski, 'A technique for pyrolysing or vaporizing samples for gas chromatographic analysis', *Nature*, **188**, 403 (1960).
9. A. Barlow, R. S. Lehrle, and J. C. Robb, 'Direct examination of polymer degradation by gas chromatography. I. Applications to polymer analysis and characterization', *Polymer*, **2**, 27 (1961).
10. T. Johns and R. A. Morris, 'New methods of pyrolysis', *Develop. Appl. Spectroscopy*, **4**, 361 (1964).
11. S. B. Martin, 'Gas chromatography. Application to the study of rapid degradative reactions in solids', *J. Chromatog.*, **2**, 272 (1959).

12. S. B. Martin and R. W. Ramstad, 'Compact two-stage gas chromatograph for flash pyrolysis studies', *Analyt. Chem.*, **33**, 982 (1961).

13. C. A. M. G. Cramers and A. I. M. Keulemans, 'Pyrolysis of volatile substances (kinetics and product studies)', *J. Gas Chromatog.*, **5**, 58 (1967).

14. F. H. A. Rummens, 'Acetate pyrolysis in the gas phase: a proposed reaction mechanism', *Rec. Trav. Chim.*, **83**, 901 (1964).

15. K. Ettre and P. F. Váradi, 'Pyrolysis-gas chromatographic technique. Effect of temperature on thermal degradation of polymers', *Analyt. Chem.*, **35**, 69 (1963).

16. S. G. Perry, 'An industrial view of pyrolysis gas chromatography', *J. Gas Chromatog.*, **5**, 77 (1967).

17. V. I. Oyama, 'Use of gas chromatography for the detection of life on Mars', *Nature*, **200**, 1058 (1963).

18. E. Reiner, 'Identification of bacterial strains by pyrolysis-gas-liquid chromatography', *Nature*, **206**, 1272 (1965).

19. E. Reiner, 'Studies on differentiation of microorganisms by pyrolysis-gas-liquid chromatography', *J. Gas Chromatog.*, **5**, 65 (1967).

20. V. I. Oyama and G. C. Carle, 'Pyrolysis gas chromatography application to life detection and chemotaxonomy', *J. Gas Chromatog.*, **5**, 151 (1967).

21. R. Fontages, G. Blandenet, and R. Queignec, 'Difficulties in the application of gas chromatography for the identification of bacteria' (in French), *Method. Phys. Anal.*, 1967, *C*63 (1968).

22. M. V. Stack, 'A review of pyrolysis gas chromatography of biological macromolecules', *Gas Chromatography 1968* (ed. C. L. A. Harbourn), Institute of Petroleum, London (1969), p. 109.

23. P. G. Simmonds, G. P. Shulman, and C. H. Stembridge, 'Organic analysis by pyrolysis-gas chromatography-mass spectrometry. A candidate experiment for the biological exploration of Mars', *J. Chromatog. Sci.*, **7**, 36 (1969).

24. J. Janak, 'Identification of the structure of non-volatile organic substances by gas chromatography of pyrolytic products', *Nature*, **185**, 684 (1960).

25. J. H. Dhont, 'Pyrolysis and gas chromatography for the detection of the benzene ring in organic compounds', *Nature*, **192**, 747 (1961).

26. J. H. Dhont, 'Identification of organic compounds from food odours by pyrolysis', *Nature*, **200**, 882 (1963).

27. J. H. Dhont, 'Identification of aliphatic alcohols by pyrolysis', *Analyst*, **89**, 71 (1964).

28. D. A. Leathard and J. H. Purnell, 'The pyrolysis of alkanes', *Ann. Rev. Phys. Chem.*, **21** (1970).

29. D. L. Fanter, J. Q. Walker, and C. J. Wolf, 'Pyrolysis-gas chromatography of hydrocarbons', *Analyt. Chem.*, **40**, 2168 (1968).

30. F. O. Rice and K. F. Herzfeld, 'The mechanism of some chain reactions', *J. Amer. Chem. Soc.*, **56**, 284 (1934).

31. A. Kossiakoff and F. O. Rice, 'Thermal decomposition of hydrocarbons, resonance stabilization and isomerization of free radicals', *J. Amer. Chem. Soc.*, **65**, 590 (1943).

32. R. Sutton and W. E. Harris, 'Optimum temperatures in pyrolysis gas chromatography', *Canad. J. Chem.*, **46**, 2623 (1968).

33. E. J. Levy and D. G. Paul, 'The application of controlled partial gas phase thermolytic dissociation to the identification of gas chromatographic effluents', *J. Gas Chromatog.*, **5**, 136 (1967).
34. J. Q. Walker and C. J. Wolf, 'Complete identification of chromatographic effluents using interrupted elution and pyrolysis-gas chromatography', *Analyt. Chem.*, **40**, 711 (1968).
35. C. J. Wolf and J. Q. Walker, 'Pyrolysis gas chromatography combined with interrupted elution for complete analysis of chromatographic effluents', in *Gas Chromatography 1968* (ed. C. L. A. Harbourn), Institute of Petroleum, London (1969), p. 385.
36. R. S. Juvet and L. P. Turner, 'Organic structure via mercury-sensitized photolysis and gas chromatography. Alcohols and esters', *Analyt. Chem.*, **37**, 1464 (1965).
37. R. S. Juvet, R. L. Tanner, and J. C. Y. Tsao, 'Photolytic degradation as a means of organic structural determination', *J. Gas Chromatog.*, **5**, 15 (1967).
38. K. A. Banks and D. D. Bills, 'Gas chromatographic identification of chlorinated insecticides based on their U.V. degradation', *J. Chromatog.*, **33**, 450 (1968).
39. O. F. Folmer and L. V. Azarraga, 'A laser pyrolysis apparatus for gas chromatography', *J. Chromatog. Sci.*, **7**, 665 (1969).
40. B. T. Guran, R. J. O'Brien, and D. H. Anderson, 'Design, construction, and use of a laser fragmentation source for gas chromatography', *Analyt. Chem.*, **42**, 115 (1970).

CHAPTER SEVEN

MEASUREMENT OF MOLECULAR WEIGHT IN GAS CHROMATOGRAPHY

An eluate cannot be identified merely from its molecular weight, but if the constituent elements are known, or known to be restricted to a certain small group, then the molecular weight can often be translated into an unambiguous empirical formula. The uncertainty attached to such a formula is strongly dependent upon the accuracy with which molecular weight measurements can be made. The method is particularly applicable to the identification of inorganic eluates, even at low levels of precision, since differences between the molecular weights of possible alternative formulae are often quite large. Organic materials are not so readily identifiable from molecular weight measurements since very different compounds may have similar, or even identical, molecular weights.

The molecular weights of some simple functions variously containing the elements C, H, O, N, and S are given in Figure 7.1. The clustering of values around 30 and 44 mass units makes it difficult to identify with any certainty alkyl derivatives with molecular weights in the regions $(30 + 14z)$, where z is the number of carbon atoms in the paraffinic chain, and similarly for simple phenyl derivatives in the regions 107 and 121 units. For example, if the measured molecular weight of a peak is 86 ± 1, i.e. close to $30 + (4 \times 14)$, the corresponding eluate could be any isomer of any one or more of the following materials: methyl propyl ketone, pentanal, hexane, N,N-dimethylpropylamine, N-methylbutylamine, pentylamine, and butenoic acid. This formidable list would undoubtedly be larger in practice, since it is based on an optimistic error in molecular weight measurement of about 1%. However, in fairness, it should be pointed out that the example has been deliberately chosen to demonstrate the sort of ambiguities which can arise, and to illustrate the limitations of identifying organic eluates from their molecular weights alone. In almost all cases additional information must be used.

The identification of an eluate is obviously much easier if its elemental constitution is known. Thus, in the example cited, it would be very useful to know whether the eluate contained nitrogen. This could readily be checked by standard methods if the sample were large enough. Alternatively, it might be possible to assign an approximate boiling point to the unknown on the basis of its retention volume (p. 47) which could be compared with those of the listed possibilities. Similarly, any background

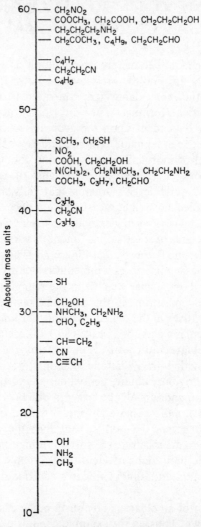

FIGURE 7.1   Molecular weights of some simple groups
containing C, H, O, N, and S

information about the sample is useful if it limits the possible eluates with a particular molecular weight.

In this chapter consideration is first given to the direct measurement of molecular weight. This involves the determination of first, the mass, W, and secondly, the number of moles, N, of a sample of eluate. The molecular weight of the eluate is then given by $M = W/N$. The mass of eluate can be determined either directly, by adsorbing the eluate and weighing it, or indirectly, by using a piezoelectric sorption detector (see p. 138). N is usually obtained by isolating the eluate from the carrier gas and measuring its pressure in a known volume at a known temperature. Then, $N = PV/RT$.

The final part of this chapter is concerned with those methods of obtaining M which depend upon measuring the density of the effluent itself without prior separation of the eluate. In particular, a full account of the gas-density balance is given.

Most of the techniques described in this chapter are simple, and relatively inexpensive. In the main, however, they are only applicable to samples of a milligram or more. Mass spectrometry is a surer method of determining molecular weights for those molecules which form abundant molecular ions, and is applicable to nanogram quantities. The coupling of GC and mass spectrometry is considered in Chapter Eleven.

7.1 MICROGRAVIMETRIC METHODS

7.1a Absolute mass detector

The weight of eluate in any peak can obviously be obtained by trapping the eluate quantitatively and weighing the trap 'before' and 'after'. Traps filled with column packing have been used in this way but the sensitivity of the method is usually low. Samples greater than about 10 mg are required, while the attainable accuracy is only of the order of ± 0.2 mg. Indeed, the percentage weight changes produced when any conventional abstractor unit is used are very small, and miniaturisation is necessary before a gravimetric detector of any reasonable sensitivity is obtained. Such a detector, consisting of a tiny sorption cell suspended from an electronic balance of high sensitivity (10^{-7} g), is shown diagrammatically in Figure 7.2. It has been used for the quantitative determination of milligram amounts of substances such as amines, alcohols, ketones, and ethers, which are irreversibly retained by a layer of concentrated sulphuric acid on the walls of the cell.[1] A smaller cell (33 mm long and 8 mm diameter), using activated charcoal as adsorbent and suitable for weighing samples of less than 0.1 mg, has also been used.[2] The weight of this complete cell is just less than 1 g, about

65% of which is charcoal. The mass detector has been used to weigh as little as $10\,\mu g$ of organic materials covering a wide range of boiling point.[3] Accuracy of about $\pm 1\%$ can be obtained.[4]

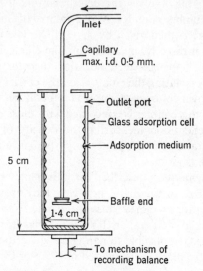

FIGURE 7.2   A simple gravimetric detector in which the eluate from the GC column is adsorbed and weighed. (From ref. 1)

7.1b   Gravimetric use of the piezoelectric sorption detector

This detector consists of a liquid-coated quartz-crystal oscillator with some means of monitoring its oscillation frequency. This frequency, f Hz, falls with increasing weight of coating, w g, according to the equation

$$f = f_0 - 2 \cdot 3 \times 10^{-6} f_0^2 \, w/A$$

where A cm² is the area of the crystal surface, and f_0 Hz is the frequency of the bare crystal. Any increase in the weight of the coating, Δw, due to solution of an eluate causes a fall in the frequency of the crystal which is directly proportional to Δw.

$$\Delta f = 2 \cdot 3 \times 10^{-6} f_0^2 \, \Delta w/A \tag{7.1}$$

Hence, a standard 9 MHz crystal with a surface area of about 1 cm² will have a sensitivity of about 200 Hz per microgram of dissolved eluate. Since frequency changes can be measured to within $0\cdot1$ Hz with standard equipment, it is clear that the detector is very sensitive. As introduced by King,[5] it was not intended to be used in this way, but rather as a selective detector (p. 152). It is evident, however, that it is potentially a very powerful gravimetric detector.

Starting from equation (7.1), it is easily shown that the weight of eluate contained in a peak is given by

$$W = \frac{AF/KV_l}{2\cdot3 \times 10^{-6} f_0^2} \int \Delta f \, \mathrm{d}t \qquad (7.2)$$

where the integral is evaluated over the period of elution, V_l is the volume of the liquid coating, F is the carrier gas flow rate at the detector, and K is the partition coefficient of the eluate between the liquid and gas phases. The latter can be obtained from the net retention volume of the peak on a conventional column when using a known volume of the liquid phase that is coating the crystal. Hence, with this column coupled to the detector, the peak-weight can in principle be obtained in a single run from the retention time of the peak on the column and the integral of the frequency change with respect to time. Substituting the typical values $A = 1\,\mathrm{cm}^2$, $F = 1\,\mathrm{ml/sec}$, $KV_l = 500\,\mathrm{ml}$, and $f_0 = 9\,\mathrm{MHz}$ into equation (7.2), and assuming that an average frequency change of $2\,\mathrm{Hz}$ is measured throughout an elution time of 50 seconds, a value of $W \approx 10^{-9}\,\mathrm{g}$ is obtained. The development of the technique to measure weights as low as this seems feasible, and would be of immense value.

7.2 MEASUREMENT OF NUMBER OF MOLES

The determination of the number of moles of an eluate is an important step towards obtaining its molecular weight. The only direct method of measuring the molar size, N, of an unknown eluate is to measure its PV product at a known temperature. Then $N = PV/RT$.

Janák[6] devised a rather complicated method of isolating eluates in the gas phase under known conditions; it involves dissolving ultra-pure carbon dioxide carrier gas in potassium hydroxide solution within a nitrometer, thus enabling the integrated volume of gaseous eluate to be measured with the gas burette. The method can be adapted in various ways to automatic operation and read-out. This is generally achieved by maintaining constant the pressure within the nitrometer by means of some feed-back mechanism which increases its volume in response to the build-up of eluate. The volume of eluate is therefore measured at constant pressure and temperature. It is claimed[7] that the apparatus shown in Figure 7.3, which is based upon volume control by a power-operated piston, enables gas volumes to be measured to within $0\cdot003\,\mathrm{ml}$.

In this technique it is necessary to establish that any particular eluate does not exceed its saturated vapour pressure within the nitrometer, and also that it is not appreciably soluble in potassium hydroxide solution. The latter requirement rules out its use with many simple organic compounds such as alcohols and acids.

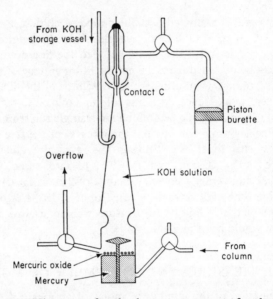

From KOH storage vessel

Contact C

Piston burette

Overflow

KOH solution

From column

Mercuric oxide

Mercury

FIGURE 7.3   Nitrometer for absolute measurement of molar size. Initially the gas space above contact C is at atmospheric pressure. The eluate, in carbon dioxide carrier gas, enters the nitrometer, and the carbon dioxide is totally absorbed in concentrated potassium hydroxide solution. The undissolved eluate rises to the top of the nitrometer, and the resulting increase in pressure causes the level of the liquid to fall until contact C is broken; a servo-mechanism then restores the pressure to atmospheric by increasing the volume of the nitrometer via the precision piston burette. The volume of the sample at atmospheric pressure, and hence its molar size, can be obtained from the total displacement of the piston. (The potassium hydroxide solution is constantly renewed from a storage bottle. A coating of HgO on the mercury surface prevents adhesion of gas bubbles.) (From ref. 7)

Phillips and Timms[8] have measured PV products by trapping the eluate, manipulating it (in a specially designed apparatus) into a known volume, and measuring its pressure with a Pearson manometer (see Figure 7.4). They were able to measure molecular weights in the range 30—200 to within 1%

from the measured PV products of milligram samples, together with the corresponding response of a gas-density detector operated with hydrogen carrier gas, i.e. in the gravimetric mode (p. 146). For materials with boiling points below 150 °C the manipulation apparatus shown in Figure 7.4 could be used at room temperature. Modified techniques were employed for less volatile and very volatile eluates, details of which are given in Table 7.1 together with some results. It can be seen that agreement between theory and experiment is very good.

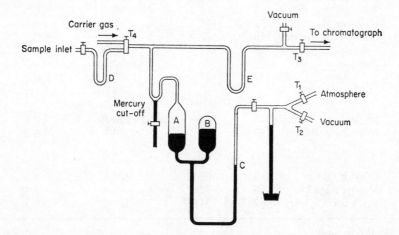

FIGURE 7.4 Apparatus incorporating a Pearson manometer for obtaining samples of known molar size by measuring the pressure exerted in a known volume (A). A pointer at B is made to point-contact the mercury surface by operating taps T_1 and T_2. This procedure is carried out (a) with A evacuated and (b) with A containing a sample at pressure p_A. The observed difference in the two readings of manometer C, Δp_C, is related to p_A by the equation: $p_A A_A = \Delta p_C A_C$, where A_A and A_C are the cross-sectional areas of A and C. Since $A_A \gg A_C$, small pressures in A can be accurately measured.
Samples are manipulated by means of traps D and E, and introduced into the chromatograph via taps T_3 and T_4. (From ref. 8)

The potential sensitivity of the method is indicated by noting that a micromole of gas exerts a pressure of approximately 20 mm Hg in a volume of 1 cc at room temperature. It is clear, therefore, that the use of an even more sensitive manometer, such as a spiral gauge or a pressure transducer, enables determination of the molar size of very small samples, provided that these can be quantitatively trapped, or separated in some other way from the carrier gas.

TABLE 7.1   Results of molecular weight determinations in GC using a gas-density detector gravimetrically with samples of known molar size obtained by PV measurements[8]

Compound	B.P. (°C)	Mol. wt.	Detd. mol. wt.	Error (%)
Apparatus I[a]				
Diethyl ether	35	74·1	74·0	−0·1
Ethyl propionate	99	102·1	102·2	+0·1
Toluene	110	92·1	92·6	+0·5
p-Xylene	138	106·2	107·1	+0·8
Tetrachloroethane	146	167·9	168·0	+0·1
Tetrachlorosilane	58	169·9	170·9	+0·6
Trisilane	53	92·3	92·4	+0·1
Tetrasilane	ca. 108	122·4	122·2	−0·2
Apparatus II[b]				
Mesitylene	165	120·2	120·1	−0·1
Dimethylaniline	193	121·2	120·8	+0·5
Apparatus III[c]				
Ethyl bromide	38	109·0	108·3	−0·6
Methyl ethyl ketone	80	72·1	72·4	+0·4
Trichloroethylene	87	131·4	131·2	−0·2

[a] See Figure 7.4.
[b] As Figure 7.4, with everything heated to 100 °C in a steam jacket.
[c] Pressure of sample, contained in a 10-ml bulb, measured with a simple manometer.

7.3   GAS-DENSITY DETECTOR

This device was first used in GC by Martin and James[9] in 1956. Ideally it gives a response which is proportional to the difference in density between the effluent gas stream and a reference stream of pure carrier gas. In turn, this difference is dependent upon the molecular weights of the eluate and carrier gas, and also upon the total weight of eluate.

The original design, shown schematically in Figure 7.5(b), was difficult to construct and to use. As a result the detector was never popular, despite its obvious merits. More recently the design has been simplified by Nerheim.[10] This form, shown in Figure 7.5(a), is commercially available and is receiving more widespread use.[11–17] (Miniature cells, $2\frac{1}{2} \times 2 \times 2$ inches, have now been introduced.[11]) It can be operated at temperatures above 200 °C, and can thus be used for fairly involatile eluates.

7.3a   Principle of operation

Basically, the gas-density detector consists of a series of interconnected channels arranged in such a way that a change in the density of the effluent gas causes a change in the flow pattern which is converted into an electrical response via two flowmeters connected in a Wheatstone bridge circuit. It should be emphasised that it is the effect of gravity upon the gas which determines the response of the detector.

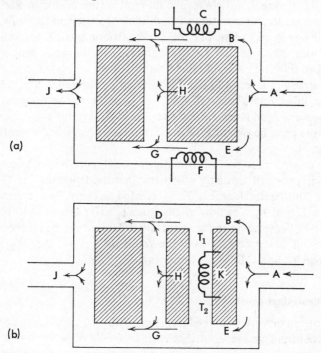

(a)

(b)

FIGURE 7.5   Two types of gas-density detector shown in vertical section. Column effluent enters at H, and reference gas at A. Mixed gases leave at J. In (a), temperatures of filaments C and F are used to detect flow changes in channels BD and EG caused by density change in DG. In (b), thermocouple junctions T_1 and T_2 detect flow of gas in the channel containing the heater filament K

The two designs shown in Figure 7.5 are in vertical section, although it is difficult to show the geometry in two dimensions. In both designs the effluent enters at H and is split into two streams HDJ and HGJ. A reference stream of the same gas as that used for carrier gas enters at A, and then passes through the channels AB and AE to join the effluent at D and G, and the mixed gases leave at J. If effluent and reference have identical densities the flow rates in BD and EG are identical. If, however, the effluent (carrier gas and eluate) has a greater density than the reference gas, then the flow in EG will tend to be restricted owing to an increase in the pressure at point G, and that in BD will be correspondingly increased. The converse will apply if the effluent has a lower density than the reference gas. The two designs differ principally in the means used to measure this flow change. In Figure 7.5(a) there is a flowmeter (thermistor or filament) in each channel; the temperature of F will tend to rise and that of C to fall. The two flowmeters form part of a Wheatstone bridge, the off-balance signal from which is fed to a recorder. The maximal sensitivity for thermistor flowmeters is obtained with a parallel-bridge circuit, while filaments require a series-bridge for optimal performance. The thermistor design is inherently about ten times as sensitive as the filament design, while the noise level is only about twice as high. Against this advantage of thermistors must be weighed the disadvantage of requiring better temperature control.

In the original design of Martin and James [Figure 7.5(b)] an additional channel is provided through which the increased pressure at G can be relieved. The heater filament K is situated between two thermocouple junctions T_1 and T_2. The flow of gas from T_2 to T_1 causes T_2 to be cooled and T_1 to be heated, and the temperature difference can be recorded. If the effluent is of lower density than the reference stream, then T_2 becomes hotter than T_1, and C becomes hotter than F.

7.3b   Theoretical considerations

The change in density produced in the effluent by the emerging peak takes place at constant pressure, so that each molecule of unknown in a given volume has taken the place of a molecule of carrier gas. The response can therefore be taken as a measure of the *mass* of the substance emerging, provided that it has a much higher molecular weight than the carrier gas— since the decrease in mass of carrier gas per unit volume can then be neglected. In this gravimetric mode (e.g. with hydrogen or helium carrier, and substances of molecular weight 100 or more) the detector is essentially a microbalance. If, on the other hand, the carrier gas has a molecular weight much larger than that of the unknown, the mass of the latter can be ignored and the gas-density signal is then a measure of the *number* of (heavy) carrier gas molecules displaced by the unknown, i.e. of the number of *moles* of sample in the emergent peak. Under these two extreme conditions, the gas-density detector can function either as a mass-meter or as a mole-meter. It is therefore very well suited to the determination of molecular weight, since $M = W/N$.

Before considering ways in which molecular weights can be obtained with the aid of a gas-density detector, it is of interest to put these intuitive ideas about the nature of its response on to an algebraic footing. This is particularly useful when eluate and carrier gas have comparable molecular weights, with the result that the detector is neither a mass-meter nor a mole-meter but something in between.

The density of a gas, ρ, is given by

$$\rho = (P/RT)\,M$$

where M is the molecular weight. Thus, at constant pressure, ρ is proportional to M. The effective molecular weight of an effluent consisting of a mixture of a carrier gas of molecular weight M_c and a sample of molecular weight M_s is

$$M = (1 - X_s)\,M_c + X_s M_s = M_c + X_s(M_s - M_c)$$

where X_s is the mole fraction of the sample. Thus, the change in density of the effluent due to the presence of sample is given by

$$\Delta\rho = (P/RT)\,[X_s(M_s - M_c)]$$

It has been found[10] that the electrical output, y, of the Nerheim design [Figure 7.5(a)] is proportional to the change in flow rate Δu, which is in turn proportional to $\Delta\rho$. Hence, we may write

$$y = k X_s(M_s - M_c)$$

where k is a given constant for a given cell, but may well alter for different carrier gases. Since X_s is always small in GC, we may make the approximation:

$$X_s = n_s/(n_s + n_c) \approx n_s/n_c$$

where n_s and n_c are the instantaneous number of moles of carrier gas and of sample per unit volume of effluent. It follows that

$$y = \frac{k n_s}{n_c}(M_s - M_c)$$

and the peak area, A, is given by

$$A = \int y\,dt = \frac{k N_s}{n_c}(M_s - M_c)$$

$$= k' W_s \left(1 - \frac{M_c}{M_s}\right) \tag{7.3}$$

$$= k' N_s(M_s - M_c) \tag{7.4}$$

where k' is effectively constant for a given carrier gas, and N_s and W_s are the total number of moles and the total mass of sample, respectively.

It is clear from equation (7.4) that, if $M_c \ll M_s$, the response of the detector for a given carrier gas is positive and proportional to W_s, the mass of sample, while if $M_c \gg M_s$ it is negative and proportional to W_s/M_s, the number of moles of the sample. In either case the response does not depend only on the value of M_s. This point is illustrated for the carrier gases hydrogen and sulphur hexafluoride in Table 7.2, which gives the ratios of the peak areas for equal mass and equimolar samples of eluates with molecular weights M_{s1} and M_{s2}. For hydrogen carrier gas, equal-mass samples give peaks whose areas differ by only 10%, even though M_{s1} and M_{s2} differ by a factor of 10. This essentially gravimetric mode of operation was used by Phillips and Timms in conjunction with PV measurements to obtain the molecular weights of Table 7.1. Similarly, for sulphur hexafluoride carrier gas, *equimolar* samples give peaks whose areas differ by only 15%.

TABLE 7.2   Calculated ratio of peak areas (A_1/A_2) for various eluates in H_2 and SF_6 carrier gases

	SF$_6$ ($M_c = 146$) $M_{s1} = 2$; $M_{s2} = 20$	H$_2$ ($M_c = 2$) $M_{s1} = 20$; $M_{s2} = 200$
A_1/A_2 for samples of equal mass	11·4	0·91
A_1/A_2 for equimolar samples	1·14	0·091

The gravimetric response of the gas-density cell with hydrogen carrier gas is in close accord with these theoretical expectations, as is demonstrated by the examples of molecular weight measurement given in Table 7.1. The response with carrier gases other than hydrogen agrees less well with theory. To ensure the highest possible sensitivity, linearity, and stability it is desirable to investigate the effect of carrier and reference flow rates on response, before carrying out measurements. It appears that in general it is best to work with a reference flow rate significantly greater than that of the effluent. The optimum flow rates will probably alter with the carrier gas used, because of the complex dependence of the electrical response of the flowmeters.

A similar relationship to equation (7.4) can be derived for the response of an ultrasonic detector[18, 19] which measures changes in the density of effluent by means of the corresponding change in the velocity of sound. However, recent work seems to indicate that currently available detectors are not suitable for molecular weight determinations.[20]

7.3c   Methods of use

There are at least four different ways of using the density balance to measure molecular weight. Methods (*i*) and (*ii*) involve only one carrier gas, while methods (*iii*) and (*iv*) involve more than one.

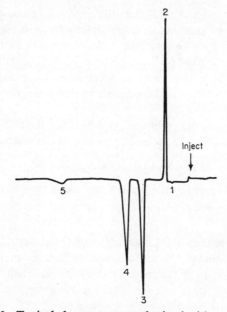

FIGURE 7.6   Typical chromatogram obtained with a gas-density balance. The carrier gas used was nitrogen.
Key: 1, N_2; 2, CH_4; 3, CO_2; 4, N_2O; 5, C_2H_6. (From ref. 11)

Method (i)

Measurement of the peak area due to a known mass of material, W_s, leads directly to M_s if the cell constant, k', is known [see equation (7.4)]. k' may be determined by measuring the area of a peak due to a known mass, W_r, of a calibration material of known molecular weight, M_r. It follows from equation (7.4) and the equivalent equation for the calibrant that

$$M_s = M_c/[1 - (A_s W_r/A_r W_s)(1 - M_c/M_r)] \qquad (7.5)$$

where A_s and A_r are the areas of the sample and calibrant peaks.

If the unknown material is available as a pure compound, it is convenient to fix W_r/W_s by adding a known mass of calibrant to a known mass of the unknown compound, and to determine A_s and A_r from a sample of the mixture. On the other hand, when dealing with a mixture of unknown materials it is necessary to determine the weight percentage of each component in the sample by a separate method. The absolute mass detector (p. 137) is useful for this purpose with relatively large samples.[3,23]

Method (ii)

It is clear from equation (7.3) that the area of a peak due to a known number of moles of sample, N_s, leads directly to M_s if the cell constant is known. The latter may be determined as described in Method (*i*) or alternatively by measuring the peak area due to a known number of moles of a calibrant, N_r. There are obvious advantages to measuring the same quantity for both sample and calibrant; in particular the same apparatus can be used.

Equation (7.3) and the equivalent equation for the calibrant lead to the following equation for the molecular weight of the unknown:

$$M_s = M_c + (A_s N_r / A_r N_s)(M_r - M_c)$$

N_s or N_r can be obtained by the gas-law methods given above in Section 7.2.[8]

Method (iii)

The main disadvantage of Methods (*i*) and (*ii*) is that they both depend upon some additional technique to measure either the masses or the molar sizes of the unknown and reference samples. If more than one carrier gas is used, molecular weights can be obtained by using the density balance alone.

In this method a reference compound is mixed with a sample of the unknown(s), and two aliquots of the mixture (assumed homogeneous) are injected into the chromatograph, the first with one carrier gas and the second with another. If M_c and M_c' are the molecular weights of the carrier gases, and the corresponding peak areas are A_s and A_s', and A_r and A_r', for sample and calibrant respectively, then it follows from equations such as (7.3) that

$$\frac{A_s(M_s - M_c')}{A_s'(M_s - M_c)} = \frac{A_r(M_r - M_c')}{A_r'(M_r - M_c)} \tag{7.6}$$

Hence,

$$M_s = \frac{M_c'(A_s/A_r)(M_r - M_c) - M_c(A_s'/A_r')(M_r - M_c')}{(A_s/A_r)(M_r - M_c) - (A_s'/A_r')(M_r - M_c')}$$

The accuracy of the method depends strongly upon the various molecular weights involved in this equation. If M_s is either very much larger or very much smaller than both M_c and M_c', the left-hand side of equation (7.6) is very insensitive to the value of M_s, and hence peak areas must be measured with extreme precision. When using hydrogen and nitrogen carrier gases to measure molecular weights above 100, an error in M_s of 4% has been quoted.[21] This error is too large for the method to be very useful in identification. However, if M_s lies between M_c and M_c', the left-hand side of equation (7.6) is more sensitive to the value of M_s, and hence small errors in peak area measurement are not so important. [See Method (iv) below.]

The composition of the mixture of sample and calibrant must be kept constant between injections, a variation of 1% being maximal for molecular weight accuracies of a few per cent. For this reason, the molecular weights of volatile liquids or unstable materials cannot be determined by this method.

Method (iv)

It is evident from equation (7.4) that when $M_s > M_c$ the detector response is positive and when $M_s < M_c$ it is negative, as shown in Figure 7.6. Hence, if the response for a given sample is measured for various carrier gases, some of lower and some of higher molecular weight, it should be possible to 'bracket' the molecular weight of the unknown. Parsons[22] studied this approach in some detail.

Figure 7.7 shows a plot of $A_s(M_r - M_c)/A_r M_r$ against M_c for a homogeneous mixture of benzene (calibrant) and a material produced from the pyrolysis of a polymer. The plot is linear, in accord with equation (7.5), and the intercept on the abscissa gives the molecular weight as 59. Infrared analysis indicated that the material was propanal, which has a molecular weight of 58. It is clear from Figure 7.1 that without this additional information the molecular weight would have been of limited use for identifying the material.

The reduction of errors by the use of carrier gases with molecular weights which bracket the unknown is apparent from Figure 7.7. Extrapolation of N_2 and CO_2 points alone leads to a value of about 62 for the molecular weight of propanal, whereas the bracketing points of N_2 and $C_2H_4F_2$ give the value 59. Using equation (7.6), instead of the graphical method, gave corresponding values of 75 with the N_2 and CO_2 points, and 59·5 with N_2 and $C_2H_4F_2$. As mentioned above, the correct value is 58. Similarly, again using equation (7.6), but with CCl_2F_2 and C_4F_8 (perfluorocyclobutane) as carrier gases, the molecular weight of carbon tetrachloride (153·8) was calculated as 155·2, i.e. to within 1%.[22]

The use of several carrier gases, rather than two, is expected to improve the accuracy of the technique, especially if extrapolation to higher molecular weights is required. Since the important variable is the *effective* molecular weight of the carrier gas, it may be possible to use an accurately metered mixture of two gases. If hydrogen and sulphur hexafluoride were to be

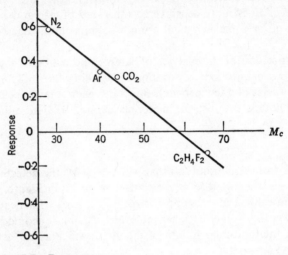

FIGURE 7.7   Response

$$\left[= \frac{A_s}{A_r}\left(\frac{M_r - M_c}{M_r}\right)\right]$$

of a gas-density detector for an unknown eluate plotted against the molecular weight of carrier gas, M_c. The intercept on the M_c axis gives the molecular weight of the unknown. (From ref. 22)

used in this way, it would then be possible continuously to vary the effective molecular weight of the carrier gas between 2 and 146.

REFERENCES

1. S. C. Bevan and S. Thorburn, 'A new quantitative integral detector for gas chromatography', *J. Chromatog.*, **11**, 301 (1963).
2. S. C. Bevan and S. Thorburn, 'Absolute mass integral detector for gas chromatography', *Chem. in Britain*, **1**, 206 (1965).
3. S. C. Bevan, T. A. Gough, and S. Thorburn, 'Quantitative analysis by mass detection', *J. Chromatog.*, **43**, 192 (1969).
4. S. C. Bevan, T. A. Gough, and S. Thorburn, 'Performance characteristics of the Brunel mass detector', *J. Chromatog.*, **42**, 336 (1969).
5. W. H. King, Jr., 'Piezoelectric sorption detector', *Analyt. Chem.*, **36**, 1735 (1964).

6. J. Janák, 'Vapour phase chromatography on zeolites', in *Vapour Phase Chromatography* (ed. D. H. Desty), Butterworths, London (1957), p. 247.
7. G. Kateman and J. A. Rijks, 'A recording chromatograph based on Janák's principle', *J. Chromatog.*, **14**, 13 (1964).
8. C. S. G. Phillips and P. L. Timms, 'Molecular weight determination with the Martin density balance', *J. Chromatog.*, **5**, 131 (1961).
9. A. J. P. Martin and A. T. James, 'Gas-liquid chromatography: the gas-density meter, a new apparatus for the detection of vapours in flowing gas streams', *Biochem. J.*, **63**, 138 (1956).
10. A. G. Nerheim, 'A gas-density detector for gas chromatography', *Analyt. Chem.*, **35**, 1640 (1963).
11. Gow-Mac Instrument Co., 100 Kings Rd, Madison, N.J., U.S.A., Technical Newsletter TNL-GADE.
12. C. L. Guillemin and F. Auricourt, 'Gas phase chromatography. Study of the Gow-Mac gas density balance. Application to quantitative analysis', *J. Gas Chromatog.*, **1** (10), 24 (1963).
13. C. L. Guillemin and F. Auricourt, 'Choice of carrier gas for the gas density balance', *J. Gas Chromatog.*, **2**, 156 (1964).
14. C. L. Guillemin, F. Auricourt, and P. Blaise, 'Operation of the Gow-Mac gas density balance', *J. Gas Chromatog.*, **4**, 338 (1966).
15. C. L. Guillemin, F. Auricourt, and P. Blaise, 'Mounting a gas density balance on a detector for calibration', *Z. Analyt. Chem.*, **227**, 260 (1967).
16. C. L. Guillemin, F. Auricourt, and J. Vermont, 'Intérêt de la balance à densité de gaz dans le contrôle industriel', *Chromatographia*, **1**, 357 (1968).
17. J. T. Walsh and D. M. Rosie, 'Studies of the gas density cell', *J. Gas Chromatog.*, **5**, 232 (1967).
18. F. W. Noble, K. Abel, and P. W. Cook, 'Performance and characteristics of an ultrasonic gas chromatograph effluent detector', *Analyt. Chem.*, **36**, 1421 (1964).
19. D. P. Lucero and A. C. Krupnick, 'Theoretical aspects of an acoustical detector possessing hydrogen specificity', *Analyt. Chem.*, **40**, 1222 (1968).
20. H. W. Grice and D. J. David, 'Performance and applications of an ultrasonic detector for gas chromatography', *J. Chromatog. Sci.*, **7**, 239 (1969).
21. A. Liberti, L. Conti, and V. Crescenzi, 'Molecular weight determination of components by gas-phase chromatography', *Nature*, **178**, 1067 (1956).
22. J. S. Parsons, 'Bracket method for molecular weight determination of pyrolysis products using gas chromatography with a gas density detector', *Analyt. Chem.*, **36**, 1849 (1964).
23. S. C. Bevan, T. A. Gough, and S. Thorburn, 'The use of gas chromatographic detectors for molecular weight determinations', *J. Chromatog.*, **44**, 241 (1969).

CHAPTER EIGHT

IDENTIFICATION FROM DETECTOR RESPONSE

Since virtually any molecular property can form the basis of a detection system, it is not surprising that some thirty types of detector have been used in GC. Although the modes of operation of these detectors vary considerably, their responses have two important features in common. First, all responses are a function of some property of the effluent. Secondly, they are also functions, normally linear, of the amount of eluate. Therefore, a recorded peak has little qualitative significance unless the amount of material producing it is known. Sometimes the amount of eluate injected *is* known, but more frequently it forms an unknown proportion of the injected sample mixture and must be determined by some separate means. The gravimetric detectors described in Chapter Seven can be used for this purpose and enable *specific* detector responses to be calculated, i.e. response per gram or per mole. The specific response of an unknown eluate can be compared with data obtained for known compounds in order to assist in its identification. These data can be determined for standard eluates on the particular detector which is to be used.

Alternatively, for detectors with linear response, published specific response data can be adapted to the actual detector being used by scaling to accord with responses obtained from known amounts of standard materials. Ideally, measurements should be made on several substances from the published data to obtain an average scaling factor. For those detectors with some element of selective response, this procedure, together with retention data, makes possible a complete identification of the eluate in some cases. A more general method of identification involves comparing the response to a given sample of any two (or more) detectors, neither of which need be gravimetric.

8.1 DETECTOR RESPONSE

8.1a Piezoelectric sorption detector

This unusual detector is based on the fact that the output from a piezo-electric material is influenced by the weight of coatings or layers on its surface. Such devices are already available commercially for use in detecting traces of water vapour, and sensitivity towards, for example, xylene can be as high as 1 part per million of carrier gas. The response of piezo-electric sorption detectors coated with liquid was discussed in the preceding chapter (p. 138), where it was shown that in principle they can be used to weigh GC peaks if the partition coefficient of the eluate between the gas phase and the liquid absorbent layer is known. Conversely, however, they can be used to obtain information about the partition coefficient for an unknown eluate, and hence they are useful qualitative detectors.[1] Of particular interest is the fact that one can modify the selectivity at will by altering the absorbent layer. GC stationary phases are useful here because of their low volatility, and it is found that relative responses correlate well with retention data. For example, if the relative response of benzene to that of cyclohexane is taken as unity with a layer of silicone oil on the piezo-electric, the corresponding value for an adsorbent layer of 1,2,3-tri-(2-cyanoethoxy)propane is 8·06. The possibility therefore arises of being able, in effect, to use the detector as a second column.

An attractive dual-detector system might be provided by two piezo-electric detectors in parallel, one of which is coated with a boiling-point stationary phase, and the other with a selective stationary phase. As with retention coincidence methods (Chapter Three), it would be as well to have several such detectors, including some coated with highly selective liquid phases, to avoid possible ambiguities.

Another possible use for this detector is as a very sensitive means of ascertaining whether an unknown eluate has reacted with a specific and fast-acting reagent; Guilbault has already carried out some static experiments along these lines.[2] This technique would essentially be an extension of conventional analysis but would be far more universal since the only requirement for a qualitative test would be that *some reaction takes place*; it would no longer be necessary for there to be a colour change, precipitation, or some similarly tangible reaction. Depending upon the eluate and reagent, we may envisage three possible types of behaviour.

(*a*) The eluate does not react. Any detector response would therefore be due entirely to partition processes, and the resulting gaussian-shaped peak would be symmetrical [Figure 8.1(a)].

(*b*) The eluate reacts to form (generally) a less volatile material. For example, concentrated sulphuric acid reacts with olefins to form alkyl sulphates. In this case the detector response would have a skew gaussian shape as shown in Figure 8.1(b). Similar shapes would result from a non-linear distribution isotherm.

(*c*) The eluate reacts to form an involatile product. For example, the reaction between sodium hydroxide and carboxylic acids yields sodium salts. The corresponding detector response would consist of a step function [Figure 8.1(c)].

For well chosen reagents, the type of detector response observed would obviously enable a great deal to be deduced about the identity of the eluate.

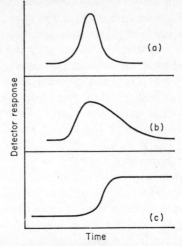

FIGURE 8.1   Hypothetical response of a piezoelectric sorption detector with the crystal coated with a reactive material: (a) reversible sorption without reaction; (b) reaction to form a less volatile product; (c) reaction to form an involatile product.

8.1b   Dielectric constant measurement

Detectors which respond to changes in the dielectric constant of the effluent enable molecules to be distinguished on the basis of electronic configuration. Early attempts to use such detectors were disappointing,[3] but recently[4, 5] a much higher sensitivity has been achieved, approaching that of many flame ionisation detectors ($\sim 10^{-10}$ g/sec). The effluent is passed through a small parallel-plate capacitance which forms part of the resonance circuit of a 70 MHz oscillator. The high sensitivity has been achieved both by the use of a small detector volume, and by using a beat technique. Although mainly used for the detection of fixed gases, and for the oxides of carbon, nitrogen, and sulphur, it has been successfully employed for mixtures containing simple hydrocarbons.

With the aid of a gravimetric companion detector, this device may prove useful in characterising substances from the relative responses obtained. For example, nitro-compounds will give a much larger specific response than most other substances. The detector should also prove useful in other dual-detector systems, since the basis of its response, namely, the polarisability and dipole moment of the eluate, are not obviously related to the response of other common detectors.

8.1c   Sensitised flame ionisation detectors

Measurement of the ionisation produced when certain molecules are burnt in a hydrogen flame is frequently used as a means of detection. There has been some measure of success in relating response of these flame ionisation detectors to effluent composition, especially for organic compounds, where the extent of ionisation reflects the number of combustible carbon atoms in the molecule.[6,7] Except for hydrocarbons, however, correlation is probably not good enough to give any reliable qualitative information about an unknown eluate.

An interesting effect with *sensitised* FID's has proved useful in routine tests for the presence of halogens or phosphorus in eluted peaks.[8] If the annular electrode of the detector is coated with sodium hydroxide, normal flame ionisation behaviour is observed unless the larger flame produced by increasing the hydrogen flow rate envelops the electrode and causes it to glow red. This brings about a slight decrease in response to most substances, but halogen-containing materials (especially bromides) and phosphorus insecticides give greatly enhanced signals (see Figure 8.2). This response enhancement for halogens, but not for phosphorus, is suppressed for electrodes coated with halides of potassium or rubidium, and hence independent tests can be made for the presence of halogen and of phosphorus.[9] On the other hand, the use of sodium chloride or phosphates of alkali metals does not produce any suppression of response.

The 'sensitised' FID is now available commercially, and utilises a solid salt jet in place of the coated electrode.[10] A caesium bromide jet has been found to give the best performance for phosphorus detection; nitrogen compounds give a response about 100 times less.[10] Similarly, in a study of the response enhancements of an FID towards several Group V elements, typical enhancement factors were as follows: P 10,000; N 100; As 30.[11] The differences between these factors allow phosphorus to be readily distinguished from nitrogen or arsenic by comparing the responses of 'sensitised' and 'normal' FID's to a given sample. The nitrogen enhancement can be increased to about 1000, while the phosphorus enhancement remains at about 10,000, by the use of a rubidium sulphate jet.[12,13]

Clearly, the general mechanism of the detector is complex, and no adequate theory of its operation exists (see e.g. ref. 14). Flame shape, temperature, and composition, electrode spacing, electrode design, background ionisation current, and sample size all seem to play important roles in determining the magnitude and even the sign of the detector response. Use has been made of the change of sign of response with background current in a sensitised flame detector for sulphur.[15] Using a potassium sulphate jet and a flame background current of about 5×10^{-9} amp, the response to microgram samples of sulphur compounds was negative, while that for hydrocarbons, phosphorus, nitrogen, and halogen compounds was positive. The magnitude of the molar response for most compounds

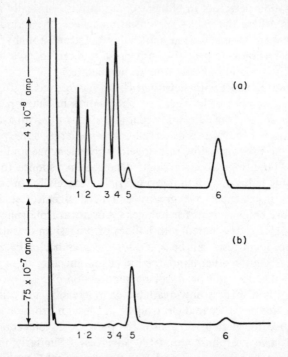

FIGURE 8.2   Chromatograms to illustrate enhanced responses obtained for chlorine-containing materials with a sensitised hydrogen flame ionisation detector: (a) normal mode; (b) electrode coated with sodium hydroxide. Samples in each case were 2 μl aliquots of a petroleum ether solution containing 1% v/v of: 1, ethyl acetate; 2, acetone; 3, trifluorotoluene; 4, fluoro-benzene; 5, chloroform; 6, monochlorobenzene. (From ref. 8)

containing one S atom was about fifty times greater than for hydro-carbons. For hydrogen sulphide the enhancement was only about 10, but for carbon disulphide it was about 100. The detector is particularly attractive for these two compounds, which are not detected by a normal FID. However, the complexity of the factors affecting the detector res-ponse is indicated by the fact that large sample sizes produce W-shaped peaks of the type sometimes found with katharometers.[15]

A modification of the detector has been described[16] which involves two FID's stacked one upon the other with a platinum wire screen between them coated with an alkali-metal salt (see Figure 8.3). In the presence of halogen (with the exception of fluorine) or phosphorus atoms, the conductivity of the upper flame increases. Hence the upper flame responds

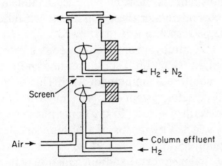

FIGURE 8.3   Schematic diagram of hydrogen double-flame ionisation detector specific for halogens and phosphorus. The platinum wire screen between the two flames is coated with an alkali-metal salt. (From ref. 16)

only to those materials which contain Cl, Br, I, or P, while the lower flame responds to all those substances normally detected by an FID. The response of the upper flame to a given element is directly proportional to the amount of that element in the eluted peak, and nanogram quantities can readily be detected. The sensitivity for phosphorus is ten times higher than that for halogens. It has been found that even fluorine can be detected if the screen is coated with a caesium salt. Optimal performance is obtained with the screen at red heat. (As with the single sensitiser flame, however, the mechanism remains obscure, despite a great deal of experimentation.[17])

The sensitivity of the detector to phosphorus can be eliminated by adding phosphorus vapour to the upper flame, thus making the detector specific for halogens. At the same time, the halogen sensitivity is increased.[17] The sensitised dual FID is insensitive to nitrogen even if the electrode is coated with a rubidium salt. However, greatly enhanced sensitivity to nitrogen can be obtained by placing a ceramic bead, coated with a rubidium salt, over the jet.[17]

A detailed study has been reported[18] of the response characteristics of a similar, coupled sodium flame-ionisation system, in which the two flames are very close together. It was found that the polarities of the two jets had a profound influence on the relative responses obtained in the sodium and sodium-free flames. In particular, if both jets were cathodes, there was an enormous increase in the normal flame ionisation response to halides and organic phosphorus compounds. Conversely, when both jets acted as anodes, the sodium flame system had the least influence on the normal flame ionisation current; under these conditions the ratio of flame ionisation to sodium flame response was maximised.

A 'sensitised' FID, employing an electrode coated with sodium sulphate, has been described, in which both sodium D-line emission and flame conductivity can be monitored simultaneously.[19] This device is an excellent example of a dual-detector system and has similarities to the flame photometric detectors described below. In the flame photometric mode, sodium D-line emission is selected by means of a 589 mμ interference filter and is detected with a photomultiplier tube. It has been shown that, depending on detector design, halogens cause sodium emission to increase or decrease. The detector used in this work was designed to give an *increased* emission in the presence of halogens, and enhancement factors ranging from hundreds to tens of thousands were obtained, in the order $Cl < Br < I$. In contrast to other work (see ref. 18), enhancement factors for phosphorus were lower than those for halogens. The detector is considerably more sensitive in its photometric than in its ionisation mode, and in the former it is responsive to sub-nanogram amounts of halogen.

This is a rapidly expanding field, and significant improvements in specificity can be expected. At the moment, however, it is sometimes difficult even to predict whether a positive or negative response will be produced. There is therefore a need for work which will shed light on the effect of jet and electrode design, and of operating conditions such as flame composition and background current, on the detector response.

8.1d Photoionisation detectors

In this detector, ions produced when the effluent is subjected to an intense source of radiation are collected and measured, as in a flame ionisation detector. The main attraction of a photoionisation device is that the energy of excitation can, in principle, be nicely controlled. Lovelock[20] was the first to report the application of the device to GC, and he has since used it as a concentration detector.[21] A diagram of the detector is shown in Figure 8.4. Helium, argon, nitrogen, and hydrogen are suitable gases for the UV source, and if a d.c. discharge is used, the pressure of gas should be below 100 mm Hg. It is, however, possible to use a helium source at atmospheric pressure if a radiofrequency discharge is used. Unless scrupulous precautions are taken to purify the gas for the discharge, the high-energy photons are removed by impurities, so that materials with the highest ionisation potentials (e.g. CO_2 and H_2O) do not give a response. In practice, therefore, the device detects only those substances which give a response on a flame ionisation detector.

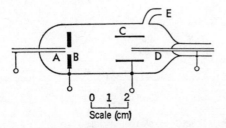

FIGURE 8.4   A simple photoionisation detector. Discharge gas enters at A, and carrier gases at D, and gases are pumped out at E. Discharge is applied between A and B, and detection takes place at electrodes C and D (anode). If a d.c. discharge is used, A is the cathode and B the anode. (From ref. 38)

A detailed study by Locke and Meloan[22] of the characteristics of a photoionisation detector has shown that the device is to some extent selective, because substances with the lowest ionisation potentials tend to give the highest response on a molar basis (see Figure 8.5). Similar data obtained by Zlatkis and co-workers[23] are shown in Table 8.1, from which it appears that the maximum UV energy obtained from the argon discharge used is between 11·6 and 12·0 eV. Photoionisation efficiencies of the detector are in the range 10^{-3}—10^{-4},[22,23] and its sensitivity is considerably higher than that of a flame ionisation detector.

All studies with photoionisation detectors reported to date[20-24] have used an internal source, as shown in Figure 8.4, where the discharge gas and effluent come into intimate contact. It would be interesting to know whether the rather poor correlation indicated by Figure 8.5 and the data of Table 8.1 would be improved if excitation were effected by an external source with a monochromator. At the present time the detector is not very much used, although it seems to have some potential as a companion detector in dual-detector systems.

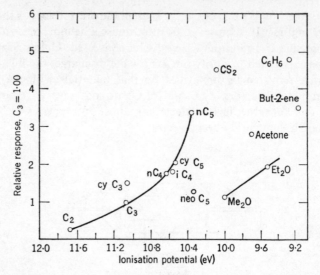

FIGURE 8.5 Observed relative molar response as a function of ionisation potential for a photoionisation detector with an argon discharge. (From ref. 22)

TABLE 8.1 Relative response of photoionisation detector to various types of sample materials[23]

Sample	Ionisation potential	Relative response
Methane	13·12	0
Ethane	11·65	0·20
Propane	11·21	1·00
n-Butane	10·80	1·29
n-Pentane	10·55	1·97
n-Hexane	10·43	2·70
n-Heptane	10·35	3·53
n-Octane	10·24	4·60
n-Nonane	10·21	5·53
n-Decane	10·19	6·68
Methyl chloride	11·29	0·91
Dichloromethane	11·35	0·86
Chloroform	11·42	0·65
Carbon tetrachloride	11·45	0·42

TABLE 8.1—*Continued*

Sample	Ionisation potential	Relative response
Methanol	10·88	0·40
Ethanol	10·60	1·93
Propan-1-ol	10·42	4·37
Butan-1-ol	10·30	6·85
Benzene	9·24	4·71
Toluene	8·82	7·85
Chlorobenzene	9·42	4·97
Methyl benzoate	9·77	4·12
Cyclopentane	10·52	2·57
Cyclopentene	10·21	1·29
Cyclohexane	10·31	4·37
Cyclohexene	9·24	5·28
Cycloheptane	—	5·71
Cycloheptene	—	8·71
Acetone	9·92	2·23
Dimethyl ether	10·52	0·68
Diethyl ether	9·72	1·09
Methyl acetate	10·32	0·77
Ethylene	10·46	1·03
Acetylene	11·42	0
Buta-1,3-diene	9·24	2·63
Hydrogen sulphide	10·47	3·17
Carbonyl sulphide	12·00	0
Carbon disulphide	10·13	4·21
2-Methylpropane	10·40	1·36
Neopentane	10·29	1·07
Vinyl chloride	10·00	3·23
Ammonia	10·25	4·08
Water	12·55	0
Carbon dioxide	14·00	0
Nitrous oxide	12·83	0
Permanent gases	—	0

8.1e Ionisation cross-section detector

The use of a chamber containing a gas and fitted with electrodes in order to detect ionising radiation goes back almost 40 years. It is only fairly recently, however, that devices of this kind employing a fixed source of radiation have been used to determine the composition of the gas within the chamber. Figure 8.6 shows a typical experimental arrangement in which the radioactive source forms the face of one of the collecting electrodes. The ionisation current is shown as a function of applied voltage

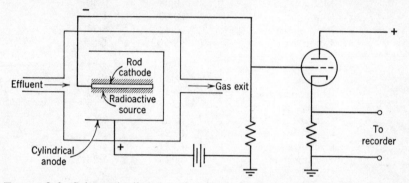

FIGURE 8.6 Schematic diagram of a simple ionisation cross-section detector and associated electrometer amplifier. Recently, miniaturisation has achieved a marked improvement in performance

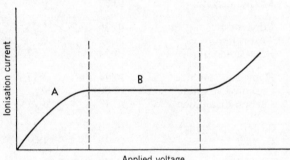

FIGURE 8.7 Current through an ionisation cross-section detector as a function of applied voltage. The height of plateau B is related in a simple way to the chemical composition of the gas in the detector

in Figure 8.7. In region A the applied voltage must be accurately controlled, since small fluctuations would result in a large change in the number of ion-pairs able to reach the electrodes before recombination took place. In region B, however, collection of ions produced by the radiation is essentially complete, and this is therefore the most useful region in which to operate the ionisation cross-section detector. At even higher applied voltages secondary ionisation is produced by the accelerated ionised particles. The actual voltage required to reach the response plateau at B depends on the separation of the electrodes; a cell of several ml volume may require 1000 V.[25] Devices suitable for GC, however, have considerably smaller volumes requiring only about 100 V for saturation for all components. In this condition, monitoring of the saturated ionisation current yields the usual gas chromatogram in which peak-heights are related to the amount and nature of the gas flowing through the cell.

It should be mentioned that the interelectrode distance should not be greater than the range of the radiation, since there would then be little change in the ionisation current measured for different gas compositions. It is therefore important that the pathlength of the ionising radiation should be much larger than the distance between electrodes, i.e. that the radiation should lose little energy while traversing the chamber. If this condition is satisfied, then the number of ion-pairs produced per second per unit pathlength is given by

$$i = kcQ$$

where c is the gas concentration in molecules per unit volume, Q is the ionisation cross-section, and k is a cell constant (for a fixed source of radiation). Q-values may be obtained from mass spectrometric measurements at low pressure.[26]

For a gaseous mixture,

$$i = (kP/RT) \sum_j x_j Q_j$$

where x_j is the mole fraction of the jth component. The area, A, of a GC peak is proportional to the extra number of ion-pairs collected from the mixture of eluate and carrier gas, and is therefore proportional to the difference in cross-sections of eluate and carrier, ΔQ, and to the total number of moles of eluate, N_s:

$$A = k'N_s \Delta Q \tag{8.1}$$

where k' is a function of flow, temperature, and pressure. Clearly, therefore, in order to obtain ΔQ, which is the characteristic parameter for any given eluate, it is necessary to know N_s (see Chapter Seven). Conversely, the detector response per mole of eluate is directly proportional to ΔQ.

Deisler et al.[25] were among the first to discuss gas analysis with such a device, and by use of an α-particle source they were able to calibrate the detector to allow determination of three-component mixtures. They found that a threshold voltage of 400 V was required for nitrogen and 70 V for hydrogen. A little earlier, a similar device had been patented which used the potentially hazardous ${}^{90}Sr/{}^{90}Y$ source,[27] and this system was later applied specifically to GC.[28] Although α-particles have higher specific ionisations than high-energy β-particles, the latter were preferred in this work because their radioactive source materials could be shielded from the carrier gas, thus avoiding possible contamination of the atmosphere. Also, because the device had a large active volume (~ 5 cc), and because the ionising particles must lose little energy in the gas phase (see above), a high-energy source was necessary. Despite the rather large volume, 100 V was sufficient to achieve saturation for all substances tested, including nitrogen.

As chromatographic techniques have improved, the need for very small detector volumes has increased so that response time can be compatible with the improved efficiencies obtainable, particularly with capillary columns. In 1963 Lovelock et al.[29] reported that they had successfully designed a cross-section detector with parallel-plate electrodes, having an effective volume of only about 0·1 ml. Since the ionising radiation in this detector travels only 1 mm, lower energies can be used for a given degree of attenuation, and so a weak β-source can be used. The use of tritium occluded in titanium or zirconium, and coated on steel electrodes, enabled a fiftyfold increase in ionisation efficiency and minimum mass detectability (10^{-9} g/sec) to be achieved, as compared with typical cross-section detectors then in use. In 1964 the same workers[30] achieved a further reduction in volume to 8 μl, suitable for capillary column use. The sensitivity of this micro-device, quoted at 3×10^{-11} g/sec, is comparable with flame ionisation techniques. At about the same time Abel[31] reported a 12 μl ionisation cross-section micro-detector which would withstand a differential pressure of 1 atmosphere without leakage.

The high sensitivity now possible with cross-section detectors does not seem to have been widely recognised, but there is little doubt that they will prove increasingly popular. Equation (8.1) indicates that their qualitative application depends upon knowing Q_j for the relevant eluates, and also upon being able to measure the specific response (per mole) for an unknown eluate. Measurement of the latter depends upon measuring N_s by the methods of Chapter Seven, while Q-values for a large number of eluates are available.[26] Also, considerable progress has been made in recent years in the theoretical calculations of Q. Good agreement has been obtained between ionisation cross-sections of atoms (relative to H) calculated from quantum mechanical considerations and as measured using both slow electrons at low gas pressures and β-particles at atmospheric pressure (Figure 8.8).[32] Also, it has been proposed[32] that the ionisation cross-section of a molecule (Q_r, relative to some arbitrary standard) can be computed by adding together those of its constituent atoms. The generality of this relationship has been challenged. Whilst it seems to hold for hydrocarbons, it does not apply to other substances, and a more general (linear) correlation is found between Q and the polarisability of the molecule.[26]

The ionisation cross-section detector is, in principle at least, a potentially useful qualitative detector. It should be possible to determine its cell constant, k', by measuring the responses of a gas-density balance and cross-section detector, connected in series, to an eluate of known Q_r; the former will measure the molar sample size, N_s (Chapter Seven), while the response of the latter will enable the cell constant to be calculated [see equation (8.1)]. This information will then enable ΔQ, and hence Q_r for an unknown eluate to be calculated.

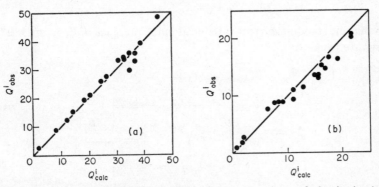

FIGURE 8.8 Plots of calculated against observed values of the ionisation cross-sections of (a) hydrocarbons and (b) non-hydrocarbons obtained from electron-impact experiments in a mass spectrometer. (From ref. 32)

8.1f Electron-capture detector

This device is one of the few detectors which have been developed with the deliberate aim of achieving a selective response for use in qualitative analysis. It responds only to those molecules which readily capture electrons; these include many oxygen- and halogen-containing materials as well as certain unsaturated compounds.

A diagram of a typical electron-capture (EC) detector is shown in Figure 8.9, from which it can be seen that it has obvious resemblances to a cross-section detector (Figure 8.6). Indeed, at high applied voltages the device acts as a cross-section detector, and the passage of an eluate E through the detector is then accompanied by an increase in current due to ionisation by the radioactive particles:

$$\beta + E \longrightarrow E^+ + e + \beta$$

At lower applied voltages, however, the ions and electrons produced in this process recombine more readily, and so the current falls towards the standing value arising only from the very low ionisation of the carrier gas, C:

$$\beta + C \longrightarrow C^+ + e + \beta$$

If, however, the eluate has a high electron affinity, the process

$$E + e \longrightarrow E^- \qquad\qquad (A)$$

becomes important. Since the drift velocity of E^- is much less than that of an electron, the process

$$E^- + C^+ \longrightarrow E + C$$

is more likely than

$$e + C^+ \longrightarrow C$$

so that the standing current is reduced in the presence of E, giving rise to a *negative* peak.

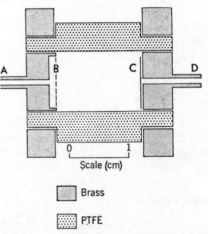

Scale (cm)

Brass

PTFE

FIGURE 8.9   An electron-capture detector. Carrier gas enters at A and passes through the brass diffuser B before leaving at D. The source of ionising radiation is at C. The entrance A acts as the detector anode and D as the cathode, so that the carrier gas moves against the flow of negative particles[38]

Whilst most molecules can be ionised with comparable ease, the ability to *capture* electrons varies considerably with the structure of the molecule and is also highly dependent upon the energy of the incident electron. At low electron energies, reactions such as (A) predominate while at higher energies dissociative capture by an eluate AB may occur:

$$e + AB \longrightarrow A\cdot + B^-$$

Operation of the EC detector with a continuous d.c. field leads to electrons with a large range of energies. In order to produce electrons with well defined energies (generally thermal) it is necessary to operate the detector in a pulsed mode in which a collecting voltage of, say, 50 V is applied for only 1 in about 100 μsec.[21,34] The addition of 5% of methane to the argon carrier gas is also recommended in order to quench super-thermal electrons.[35]

One disadvantage of the normal pulsed EC detector is that the response becomes non-linear for reductions of the standing current by more than about 1%. In qualitative work, such behaviour is unacceptable since specific and relative responses cannot be measured with any certainty. Originally it was believed that EC detectors obeyed a Beer-type law of electron absorption. Recently[36] it has been shown that this is not so, but that instead the function $(i_0 - i)/i$ varies linearly with eluate concentration where i_0 is the current flowing with only carrier gas and i is that in the presence of eluate. Since detector response is equivalent to $(i_0 - i)$, an analogue converter must be used to obtain the linear function.[36] For o-dichlorobenzene, such a converter has been shown to give a linear response from a pulsed EC detector over a range of concentrations of four orders of magnitude. This device should prove very useful in many analyses, particularly those in which the EC detector is used as a companion detector in a dual-detector system.

The energy of the electrons in an EC detector can be varied by applying a radiofrequency field, and in this way Lovelock and co-workers[37] have been able to obtain the electron absorption coefficient of an eluate as a function of electron energy (the electron attachment spectrum, EAS). Eventually it is hoped to be able to obtain the EAS of an eluate as it emerges from the chromatograph, and the authors suggest that the technique may assist in the identification of an unknown molecule. The method is, as yet, in its early stages of development, and if it is to be of use in qualitative analysis it will be necessary to show that there is a significant difference between the EA spectra of chemically similar materials.

The EC detector is frequently used in qualitative organic analysis to detect materials containing O, S, P, and halogens, the response to the latter being in the order $I > Br > Cl > F$.[38] Alkanes give no response but certain unsaturated materials, such as polycyclic hydrocarbons, are readily detected. Table 8.2 shows relative sensitivities for a number of such compounds for both EC and FID detectors.[39] The large difference between the EC sensitivities for 1,2-benzanthracene and chrysene is remarkable since these hydrocarbons have very similar structures. It should be noted that their FID sensitivities are very similar, as would be expected. The use of published values of EC responses is of limited value since they depend on a number of factors, some of which are indeterminate. Flow rate has been shown to be an important factor.[40]

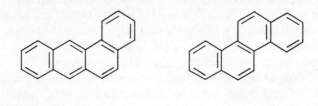

1,2- Benzanthracene   Chrysene

TABLE 8.2   Responses per unit weight for some poly-cyclic hydrocarbons relative to benzo(*mno*)fluoranthene.[39]

Eluate	FID	EC	EC/FID
Anthracene	0·75	0·06	20·2
Fluoranthene	0·76	0·09	32·5
Pyrene	0·84	0·40	124·3
1,2-Benzofluorene	1·00	0·02	5·5
3-Methylpyrene	0·97	0·27	69·2
Benzo[*mno*]fluoranthene	1·00	1·00	250·0
1,2-Benzanthracene	0·90	0·87	267·2
Chrysene	0·85	0·005	1·5
3,4-Benzopyrene	1·53	2·15	343·5
1,2-Benzopyrene	0·66	0·75	310·0
3,4-Benzofluoranthene	1·01	0·67	180·2
Perylene	1·80	0·01	1·5

The EC detector is frequently used in pesticide analysis. The early studies of halogenated materials[41] have been extended to phosphates,[42–45] and there has been some success in correlating the structure of phosphate pesticides with EC response. For example, Cooke et al.,[42] in a study of the Systox and Parathion families of phosphate pesticides, conclude that the order of the EC response to certain functional groups is as follows:

$$O=P—O < S=P—O < O=P—S < S=P—S$$

In this work comparisons were made between near-maximal sensitivities, determined by response measurements at various voltages, for a variety of pesticides (in a solution of benzene) on EC detectors operating in the d.c.

mode. There are exceptions from the order of response stated above, and therefore great care should be taken in the interpretation of such techniques. Indeed, this caution should be exercised in all uses of the EC detector, particularly in the d.c. mode, since its operation can be greatly affected by small traces of impurity, and also by relatively slight changes in its operating parameters.

Use of the EC detector can be extended to those materials which do not normally respond by forming suitable derivatives. Landowne and Lipsky appear to have been the first to suggest this approach. They discovered that monochloroacetates of sterols[46] gave markedly larger responses than the parent compounds, and this technique has recently been extended to phenols. The transition-metal complexes of tri- and hexa-fluoroacetyl-acetones[47,48] have recently been much used for quantitative inorganic analysis by GC.[49,50] Ross has made a study of the chromium and aluminium complexes with these fluorinated ligands, as well as with acetylacetone, and concludes that the metal atom has a significant effect on the observed EC response.[50,51] Although the greatest use of the EC detector is probably in a dual-detector system, its specific response behaviour may sometimes be of use by itself. For example, the detection of traces of pesticide in a large excess of hydrocarbon solvent is readily achieved since the EC detector does not respond to the latter. However, in general the detector is most useful in qualitative analysis when it is employed in connection with another detector which enables *relative* response factors to be determined. This theme is well illustrated in the final column of Table 8.2, and is further expanded below (p. 180).

8.1g   The Beilstein detector

Flame photometric devices, of varying complexity, have come to the fore in recent years as highly specific detectors. Simple flame tests, of course, have been used for many years in qualitative analysis, and one of them, the Beilstein test, has been adapted for use with GC. This well known test depends upon the emission of an intense green coloration when compounds containing halogen (except fluorine) are fed to a Bunsen flame contaminated with copper atoms. In GC the test can be carried out by using a glass burner with a helical copper wire held in its flame. For visual observation, this device has a detection limit of about 0·2 mg of chlorine, although this is to some extent subjective.[52] Much greater sensitivity can be achieved[53] by passing the column effluent through a copper 30-mesh screen thimble held in the outer cone of a cool Bunsen flame. In this method, 5 μg of chlorine in a 30-second peak can be detected by eye in a lighted room. It should be noted that samples greater than about 1 mg lead to a persistence of the green colour long after the GC peak has passed.

Monkman and Dubois[52] have described a more convenient, automated version of the technique, which uses a commercial flame photometer, with a 520 mμ interference filter. In this, the burner top is fitted with a copper baffle, and the GC effluent from a katharometer is fed into the flame by means of a number of hypodermic needles. This is a good example of a dual-detector system. Peaks containing halogen are readily determined, since only they give a photometric response. However, the detection limit achieved is poorer than with the visual method, being about 3 mg of chlorine. Despite the obvious attractions of this technique, it is only very recently that anybody has bothered to develop it for use at the microgram level.[54] This is probably because other, more generalised, photometric techniques have recently been intensively studied in connection with GC.

8.1h   Flame photometric detectors

The early work in this field was done by Grant,[55,56] who monitored the total light emission from a hydrocarbon flame in a dual-detector arrangement with a katharometer. Although compound type (e.g. aromatic, paraffinic, etc.) could be determined from the relative response of the two detectors, the device was non-selective in that emitted light from a wide range of wavelengths was detected.

More recently, Juvet and Durbin[57,58] have achieved a highly specific mode of operation by using a spectroscope. For example, with their instrument, chromium compounds can be readily distinguished from those of iron, even when both of these elements occur in a single peak. In favourable cases as little as 10^{-11} mole of inorganic materials can be detected. The sensitivity for organic compounds is similar to that of katharometer detection ($\sim 10^{-7}$ mole). Furthermore, a similar detector, though a factor of 10 less sensitive, can be constructed for as little as £200.[59] It consists of a commercial oxy-hydrogen burner, the emission spectrum of which is analysed by a diffraction spectroscope with photomultiplier detection. The wavelength range 355—625 mμ is covered with a bandwidth at half-intensity of 6 mμ. This device can also operate non-selectively, in a manner similar to that used by Grant.

Specific detection of silicon has been carried out spectroscopically by monitoring 251·6 mμ emission in an oxy-acetylene flame, or absorption in a nitrous oxide–acetylene flame. Ratios of the responses of flame photometer and katharometer enable the number of silicon atoms per molecule to be estimated.[60]

A monochromator is probably desirable for observation of the narrow emission lines due to atoms. However, for many organic applications, interference filters which transmit reasonably closely to the centre of the broad emission bands of diatomic species such as C_2 and CH will suffice. To date there have been no reports of fast scanning of GC effluent emission. However, Figure 8.10 shows the emission spectra of pentane and

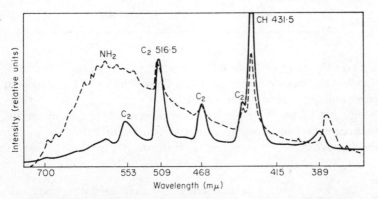

FIGURE 8.10   Hydrogen–air flame emission spectra of pentane (full line) and triethylamine (broken line). Intensity measurements at chosen wavelengths form the basis of one method of peak identification with flame photometric techniques. (From ref. 61)

triethylamine, obtained by conventional hydrogen–air flame photometry.[61] It should be noted that the low-intensity emission at 388 mμ, due to CN, is very much lower than that observed in the microwave stimulated spectra discussed in section 8.11.[62] Bandwidths of most interference filters are less than 10 mμ, at half peak transmission, i.e. much narrower than the bands of Figure 8.10. Wavelengths of interference filters within the range of Figure 8.10 used by Braman[61] are given in Table 8.3. Using flame ionisation detection on the same flame as a reference, he made use of the selectivity afforded by monitoring flame emission of GC eluates at various wavelengths to aid compound characterisation. The utility of the method is apparent from Table 8.4.

It can be seen that progressive chlorination of methane tends to increase FE/FI ratios at all wavelengths, particularly at 515 and 589 mμ, while the chlorination of benzene leads to a reduction in these ratios. Similar trends occur in the ratio of C_2 (515 mμ) to CH (420 mμ) emission. Replacement of the OH group in propan-2-ol by Br increases the FE/FI ratio at 589 mμ,

TABLE 8.3   Useful interference filters for flame photometry[61,63]

Filter wavelength (mμ)	Corresponding emission
380	CN 388·3 mμ
394	S$_2$
420	CH 431·5 mμ
515	C$_2$ 516·5 mμ
526	HPO
589	NH$_2$ broad

but leaves the ratio of C$_2$ to CH emission little changed. Substitution of CN, however, nearly doubles this ratio. The difference in C$_2$ emission from acetone and the propanols is noteworthy.

Brody and Chaney[63] developed an interference filter system specifically for the detection of phosphorus and sulphur at or below the nanogram level. The experimental arrangement is shown in Figure 8.11. Nitrogen is

TABLE 8.4   Relative flame emission and flame ionisation responses from a single flame as aids to compound characterisation[61]

Compound	Ratio of FE/FI (Filter wavelength (mμ))			Ratio of 515 mμ (C$_2$) to 420 mμ (CH) emission
	589	515	420	
CH$_3$OH	0·07	0·23	0·22	1·0
CH$_3$NO$_2$	0·58	1·05	0·33	3·2
CH$_3$CN	0·56	—	0·38	—
CH$_3$Cl	0·06	0·40	0·20	2·0
CH$_2$Cl$_2$	0·28	0·90	0·31	2·9
CHCl$_3$	0·65	1·6	0·44	4·1
CCl$_4$	0·72	2·3	0·59	4·6
n-C$_3$H$_7$OH	0·037	0·22	0·14	1·6
i-C$_3$H$_7$OH	0·029	0·24	0·19	1·3
CH$_3$COCH$_3$	0·047	0·11	0·20	0·6
CH$_3$CH(CN)CH$_3$	0·25	0·60	0·26	2·3
CH$_3$CH(Br)CH$_3$	0·25	0·37	0·27	1·4
n-C$_5$H$_{12}$	0·084	—	0·48	—
cyclo-C$_6$H$_{12}$	0·11	0·68	0·48	1·4
C$_6$H$_6$	0·082	1·12	0·37	3·0
C$_6$H$_5$Cl	0·025	0·21	0·11	1·9
C$_6$H$_5$Br	0·027	0·11	0·085	1·3

used as carrier, and is pre-mixed with oxygen at the column exit to give an air-like mixture. Hydrogen is added at the burner to give a hydrogen-rich flame. One unfortunate feature of this device is that a large excess of solvent may extinguish this sort of flame, which can only be re-ignited if the hydrogen flow is temporarily reduced.

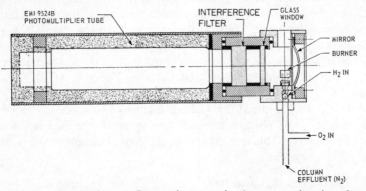

FIGURE 8.11 Oxy-hydrogen flame photometric detector using interference filters to enable specific detection of phosphorus from its emission at 526 mμ and sulphur at 394 mμ. (From ref. 63)

An essential feature of the detector is that background emission is shielded from the optical system by a housing which surrounds the flame. In the presence of P or S, emission occurs above this housing, and the light gathered by the mirror passes through the glass window and an interference filter to the photocathode of an end-on type photomultiplier. Interference filters of 526 mμ for P and 394 mμ for S are used. It will be seen from Table 8.3 and Figure 8.10 that emission from C_2 for eluates such as hydrocarbons would interfere with the phosphorus detection. Such emission remains shielded, however, presumably because the reactions leading to C_2 formation are much faster than those responsible for the phosphorus emission. Thus, a much higher degree of specificity and sensitivity is achieved than would be possible with an unshielded flame. Sub-nanogram quantities of insecticides containing phosphorus can be detected (see e.g. ref. 64). For sulphur the detection limit is about a hundred times higher. The response to phosphorus is linear with sample size, while the response to sulphur is roughly proportional to the square of the concentration. Thus, at the μg level, sensitivities for P and S are very similar. It seems that the emission from phosphorus compounds at 526 mμ is due to HPO, while that at 394 mμ is part of the band system of S_2, a fact which probably accounts for the observed concentration dependence of the sulphur response. A developed form of the detector employing two photomulti-plier tubes and capable of simultaneous detection of P and S has recently been described.[65] This system has an advantage over many other dual-detector systems in that it does not involve stream-splitting. Both the single and dual photometric detectors are available commercially.

It is particularly interesting to note that the ratio of the P response to the square-root of the S response can be used to determine the atomic ratio of phosphorus to sulphur. Thus, in Bowman and Beroza's work,[65] $R_P/\sqrt{R_S}$ had values (in arbitrary units) in the range 5·2—6·0 for PS compounds such as Parathion ($C_{10}H_{14}NO_5PS$), values about half of this, 2·8—3·3, for PS_2 compounds such as Malathion ($C_{10}H_{19}O_6PS_2$) or Ethion ($C_9H_{22}O_4P_2S_4$), and values of about one-third of the PS values, i.e. 1·7—2·3, for PS_3 com-pounds such as Phorate ($C_7H_{17}O_2PS_3$). Use could be made of this finding with only a single photometric detector by using the two interference filters in sequence. This technique is an attractive example of the power of dual-detector systems.

8.1i   Plasma emission detectors

Spectroscopic analysis of the light emitted by excited species formed from eluates in flames and plasmas is a highly sensitive method of elemental analysis. Simple flame-photometric techniques were considered in the previous section. We now consider other methods of identification based upon plasma emission, particularly in connection with microwave and electric discharges in effluents.

Excited species are formed in discharges by the reactions of high-energy electrons which have been accelerated in an electric field. For example, emission spectra may be obtained at room temperature from eluates fed to microwave-stimulated plasmas. A block diagram of an apparatus used for this purpose by McCormack et al.[62] is shown in Figure 8.12. With this set-up, high sensitivities were obtained, considerably in excess of those obtain-able with simple flame-photometric techniques. The discharge was struck in a quartz tube connected to the column exit. The diameter of this tube was critical. Using helium carrier gas, stable discharges could only be achieved at pressures of a few mm Hg, and in tubes with a diameter greater than 10 mm. With argon carrier gas, however, the discharge could be struck at atmospheric pressure, and best results were obtained in a tube of about 1 mm bore. This reduced volume of the detector system is clearly advantageous. Using this arrangement, bands due to C_2, P, CS, CCl, etc. were detected at the appropriate wavelengths, and in certain cases the occurrence of them could be related to the structure of the eluate. How-ever, care must be taken in the interpretation of such experiments, since 'flame' reactions may produce atomic groupings which do not exist in the parent material.

Figure 8.13 illustrates the specificity of the microwave system. Sensitivity was reported to be about 10^{-16} g/sec for n-hexane, using the CN emission

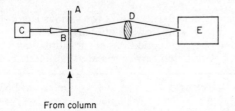

From column

FIGURE 8.12 Schematic diagram of apparatus for detecting emission spectra in the ultraviolet region from eluates excited in a 2450 MHz microwave discharge. Effluent passes through the quartz discharge tube A which is placed in the microwave cavity B, in communication with the microwave source C. Emitted radiation is focused by the quartz lens D and analysed by the scanning spectrophotometer E. (From ref. 62)

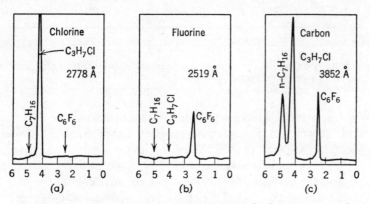

FIGURE 8.13 Chromatograms of a mixture of n-heptane, propyl chloride, and hexafluorobenzene obtained with the apparatus of Figure 8.12. Each chromatogram was obtained with the spectrophotometer locked on one wavelength. This enabled specific detection of Cl, F, and C, as indicated. (From ref. 62)

at 388·3 mμ. This was the most intense band found for all organic compounds, despite the fact that notionally no nitrogen was present in the discharge, and so must have resulted from impurities.[62] The importance of the CN band has been confirmed by other workers.[66] This is an important difference between the spectra emitted by flames and discharges. Similarly, in the microwave system, strong OH emission was observed in the UV, even for hydrocarbon eluates. Although purification of the argon carrier may eliminate this effect, it is possible that the oxygen comes from the discharge tube.

TABLE 8.5   Sensitivities and selectivities for various elements, using emission from a microwave discharge detector[62]
(Values quoted are for typical conditions of slit width, flow rate, and microwave power, with argon carrier gas at atmospheric pressure)

Compound	Wavelength (mμ)	Assignment	Detection limit[a] (g/sec)	Selectivity[b]
n-C_6H_{14}	388·3	CN	2×10^{-16}	—
	516·5	C_2	3×10^{-14}	—
C_6F_6	561·1	?	3×10^{-12}	10
	251·6	Si[c]	5×10^{-10}	20
$CHCl_3$	278·8	CCl	2×10^{-12}	20
$CHBr_3$	298·5	?	2×10^{-7}	10
CH_3I	206·2	I	7×10^{-14}	10^4
$(C_2H_5O)_3PO$	253·6	P	1×10^{-11}	10^2
CS_2	257·6	CS	1×10^{-9}	10^2

[a] Detection limits refer to the hetero-element, except for n-hexane where they refer to the compound.
[b] Selectivity is defined as the ratio of the response to the named compound to the response to n-hexane at that wavelength.
[c] The silicon is presumed to have come from the quartz tube, via SiF_4.

With the aid of a grating monochromator the microwave detector can be made specific for characteristic emission bands. Table 8.5 shows the lines used by McCormack et al., together with the sensitivity achieved, and also the selectivity, defined as the ratio of the response to the selected compound and the response to n-hexane at that wavelength. The detection limits and selectivities quoted are only representative, and are functions of conditions of flow, monochromator slit-width, and microwave power level. Also, even for a given element, there can be pronounced variation in sensitivity for different compounds. Thus, Bache and Lisk,[67] who used this detector to determine organophosphorus insecticide residues by monitoring the sharp 253·6 mμ P emission line, found sensitivities in the range 10^{-11}—10^{-12} g/sec, in reasonable accord with the figures of Table 8.5. However, in a study of iodinated herbicide residues[68] with the same detector, the detection limit of about 5×10^{-10} g/sec was considerably poorer than that indicated in Table 8.5.

The possibility of operating the discharge below atmospheric pressure has already been mentioned. Against such a procedure it can be argued that the number of excited species will fall owing to a reduced concentration of eluate. On the other hand, electrons will have a longer mean-free-path, and hence attain higher velocities in the electric field, and there will be a reduced rate of recombination. It appears that these conflicting factors result in an optimum working pressure, which for argon is at about 200 mm Hg. At this pressure Bache and Lisk found an order of magnitude improvement in sensitivity ($\sim 5 \times 10^{-13}$ g/sec) and selectivity ($\sim 10^3$) for organophosphorus insecticides as compared with atmospheric pressure. This allowed emission spectra to be obtained at the 1 part in 10^9 level.[69] It should be mentioned that the discharge was not started until a few minutes after injection, in order to avoid extinction of the discharge by the large amount of solvent in the sample, and to prevent carbon formation in the discharge tube.

Several modifications of the microwave discharge detector have been described.[70-72] Helium carrier has been used, for which, as mentioned above, it is essential to work at reduced pressures. Chloro- and bromo-methyldimethylsilyl derivatives of pesticides in the 10^{-7}—10^{-9} g range have been determined in this way[71,72] by monitoring the Cl emission line at 479·4 mμ and the Br line at 478·6 mμ. A good monochromator is required to distinguish these two lines. Comparison with an electron-capture detector showed that the low-pressure helium microwave device was usually less sensitive by a factor of about 4. Of course, the EC detector did not give as much qualitative information.

Braman and Dynako[73] have made a detailed study of emission from a d.c. discharge, which they point out has a built-in advantage over the microwave version in that it is smaller, and needs a simpler and smaller power supply. Using a grating monochromator and a wide-range, quartz-windowed, end-on photomultiplier, with a very low dark-current (which was minimised by cooling in solid carbon dioxide), they obtained detection limits for 2-bromopropane at various wavelengths as indicated in Table 8.6. As with the microwave system, it is interesting that the CN band offers the best sensitivity! Also shown in Table 8.6 are the detection limits for various halogenated hydrocarbons, using halogen atom emission lines. It is difficult to compare directly the results of Tables 8.5 and 8.6 since different spectral lines are involved. However, it is interesting that the detection limits of F and Cl are similar to those found with the atmospheric pressure argon microwave device, the I value is somewhat worse, but the Br figure is substantially better. Also, detection limits for the d.c. discharge are a hundred or more times lower than those found with the low-pressure helium microwave detector.[71] Very much greater selectivity for I is possible using the 206·2 mμ line, as used in the microwave study (Table 8.5). Otherwise, selectivities obtained in the two approaches are very similar, and the sensitivities are, by any standards, excellent. Using an interference filter of 431·5 mμ for the 431·2 mμ CH band, instead of a spectrograph, the detection limit for n-hexane with the d.c. detector was found to vary from about 5×10^{-14} g/sec to 10^{-12} g/sec, depending on the power level and the flow rate.

In summary, plasma emission spectroscopy as applied to GC effluents is a highly sensitive means of elemental analysis. Experimentally, there are some advantages to using a d.c. discharge.

TABLE 8.6   Detection limits for halogenated hydrocarbons at various wavelengths, using a d.c. discharge detector[73]

Compound	Wavelength (mμ)	Assignment	Detection limit$^a \times 10^{14}$ (g of compd./sec)
$CH_3CH(Br)CH_3$	387·5	CN	3
	431·2	CH	80
	516·5	C_2	200
	734·8	Br	2×10^3
C_6H_5F	690·2	F	100
	685·6	F	80
$C_2H_4Cl_2$	725·7	Cl	100
C_3H_7I	546·5	I	200
	608·2	I	800

^{a} These detection limits refer to the compound and depend very much upon conditions.

8.1j   Coulometric detectors

Several elements can be determined absolutely, and in some cases specifically, by means of coulometric detectors. These devices therefore enable limited elemental and gravimetric analysis to be carried out. Hersch, the inventor of the well known cell for determining oxygen, has reviewed the whole field of measurement by electrochemical means.[74]

The simplest electrochemical detector which can be envisaged consists of a cell through which the GC effluent is passed. Any reaction between eluate and electrolyte ions changes the potential difference of the cell owing to removal of these ions. In coulometric detectors suitable for use with GC this basic system has been modified so that the concentration of electrolyte is maintained constant by restoring any loss due to reaction. The eluate is said to be *titrated* against the electrolyte ions.

A schematic diagram of one arrangement which enables this to be done is shown in Figure 8.14. Effluent is passed into the cell electrolyte F (say Ag^+), which is rapidly stirred for good mixing. The potential difference between the sensor electrode C and the reference electrode D is biased so that in the absence of reaction there is zero output from the null-point amplifier E. When reaction occurs, the potential of C is lowered, as a result

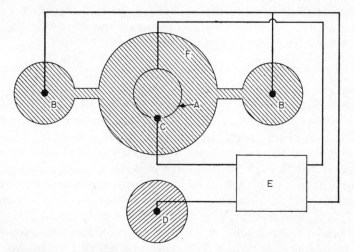

FIGURE 8.14   Schematic diagram of a suitable arrangement for coulometric titration of eluates. (A) anode; (B) cathodes; (C) sensor electrode; (D) reference electrode; (E) null point amplifier; (F) electrolyte. The current output from E traces the GC peak

of which the amplifier passes current between A and B, thereby generating ions (e.g. Ag^+ if anode A is silver metal) until the electrolyte concentration is restored, at which point the amplifier output is again zeroed. The current output of E traces the envelope of a normal GC peak, while the total number of coulombs passed (proportional to peak area) is equivalent to the number of molecules of eluate in the peak.

Originally this type of detector was devised for chloride estimation by titration against Ag^+.[75] More recently sulphur (as SO_2), down to 10^{-8} g, and mercaptans have been determined by titration against I^-.[76-78] The sulphur content of eluates is converted into SO_2 by combustion in oxygen at about 1000 °C. At the same time, free halogens and oxides of nitrogen may be produced, and since these products also react with I^- they therefore interfere with the estimation of sulphur. An existing commercial instrument (Titrilog, C.E.C.), which involves the titration of oxidants such as SO_2 against Br^-, has been modified for use with GC.[79]

Nitrogen content of eluates can be detected coulometrically by conversion into ammonia (using a Ni–MgO catalyst) and titration against hydrogen ion.[80] Interference from most acids is avoided since they are removed by the MgO; no interfering basic materials seem to be produced from organic materials under these conditions.

Coulometric detectors have proved very useful in the detection of trace, sulphur-containing components in petroleum,[77] and also in the identification of low levels ($\sim$0·04 p.p.m.) of pesticides.[81]

8.1k   Electrolytic conductivity detector

Changes in the electrical conductivity of water, due to the solution of simple molecules produced from eluates, form the basis of this detector, which is now available commercially.[82] The eluate may be converted into suitable, highly ionic molecules by a number of processes such as pyrolysis, hydrogenolysis, etc. For example, hydrogenolysis of halogenated compounds at 800 °C yields hydrogen halides which produce relatively large changes in conductivity, so that the device is virtually specific to halogens. Similarly, specific detection of nitrogen in an eluate is achieved by conversion into ammonia on a nickel catalyst, followed by absorption of any acids produced by $Sr(OH)_2$.[83] An electrolytic conductivity detector for the absolute measurement of CO_2 down to levels of 1 μg has been described.[84] CO_2 is absorbed in an electrolyte of sodium hydroxide.

8.2 DUAL-DETECTOR SYSTEMS

Two heads are better than one, and so are two detectors. Almost all detectors give responses which are functions of sample size as well as of some molecular property. Therefore, the response from a single detector is not very useful for purposes of identification unless the sample size of the eluate is also known. One way of obtaining the latter is to measure it with a second, gravimetric detector connected in series or in parallel to the first. This arrangement, which yields the response of the first detector per mole (or per gram) of eluate, is just one example of the many possible dual-detector systems. Other systems may involve two detectors, neither of which is gravimetric, while others may consist of a single detector with two modes of response (e.g. flame emission and flame ionisation). Since there are, in theory, so many possible dual-detector systems, it is desirable to consider those characteristics which are essential to a good system.

(a) Relative responses, i.e. the ratio of the individual detector responses to the same eluate, must vary significantly with the identity of the eluate.

An ideal situation is when the relative response is the same for all members of a given class.

(*b*) Both detectors must exhibit a linear response curve so that the relative response for a given material does not vary with its sample size, and is therefore characteristic of the material.

(*c*) Both detectors should be reproducible and easy to calibrate so that the determination of the relative response of the system towards a single known substance enables those for all others to be calculated; it is unrealistic to expect relative responses to be independent of the actual instruments used.

Figure 8.15 shows a dual chromatogram obtained with microcoulometer and katharometer detection,[77] and many other dual-detector systems have been described (e.g. flame emissivity and katharometer,[55,56] flame photometric and katharometer,[60] flame photometric and flame ionisation,[19] and flame ionisation and KCl-sensitised flame ionisation[9]). Systems involving flame-photometric detectors are potentially very useful in eluate

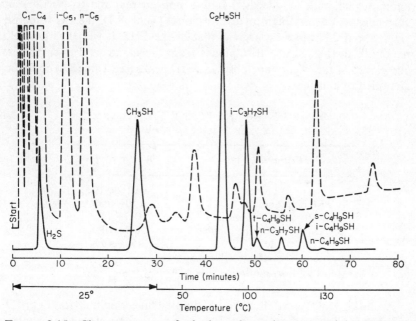

FIGURE 8.15   Chromatograms of a hydrocarbon mixture containing H_2S and mercaptans: broken trace, katharometer detection; full trace, microcoulometric titration against Ag^+ (approximately $\times 100$ sensitivity of katharometer for sulphur). (From ref. 77)

identification since the spectral range covered can be varied merely by changing an interference filter. If two flame-photometric detectors are used together it should be possible, by suitable selection of interference filters, to distinguish between any two eluates which have substantially different emission spectra. Plasma emission detectors also have similar properties.

Most applications of relative response techniques have, however, involved EC detectors because of the large degree of specificity shown by such detectors. It should be noted that at high concentrations of eluate the EC detector is non-linear, and must be used in conjunction with an analogue converter[36] (see p. 167) if it is to meet requirement (c) above. Companion detectors used with EC have included the photoionisation detector[21] and the argon β-ionisation detector[38] (see Figure 8.16), but the most frequent companion is the FID. A series arrangement of EC and FID has been used to detect polynuclear aromatics[85] (cf. Table 8.2) in peaks much overlapped in the FID chromatogram. FID and EC have been used in parallel to identify the volatile sulphides in garlic, with the aid of numerous sulphide standards.[86] It was noticed that the trisulphides in particular had a very high relative response (Table 8.7), the EC response being about 250 times as large as that of the FID. It was thus possible readily to distinguish the trisulphides from the mono- and di-sulphides, and eleven of the twenty-one observed FID peaks (see Table 8.8) could be identified by this means.

TABLE 8.7   Relative EC and FID responses
for organic sulphur compounds[86]

Compound type	Range of ϕ^a
Monosulphides	0·01—0·10
Mercaptans	0·10—0·50
Saturated disulphides	0·6—4·0
Unsaturated disulphides	8·0—20·0
Trisulphides	150—400

^{a} ϕ = Electron capture response/flame ionisation response.

Zielinski et al.[87] have described a simple effluent splitter for EC/FID, suitable for any dual detector, which uses a micrometer valve to control the splitting ratio between the two detectors. As described, the splitter has the severe disadvantage that retention times as measured from EC and FID responses differ by as much as 20 seconds. Suitable modifications to the 'plumbing' would, presumably, eliminate this. The EC/FID ratios obtained with the splitter for *o*-, *m*-, and *p*-chloronitrobenzenes were 9·0, 9·5, and 7·6. Hence, even though the *m*- and *p*-isomers had virtually identical retentions on the QF1-fluorosilicone column used, they could be readily distinguished on the basis of their EC/FID ratios.

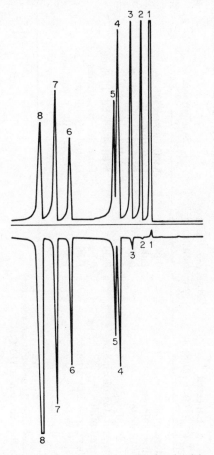

FIGURE 8.16 Dual chromatogram obtained with argon *β*-ray ionisation (upper trace) and electron capture (lower trace) detectors. Peak-height ratios assist in the identification of peaks. Chromatograms obtained with 1 *μ*g of a mixture of: 1, cyclohexane; 2, fluorobenzene; 3, chlorobenzene; 4, *m*- and *p*-dichlorobenzene; 5, *o*-dichlorobenzene; 6, 1,3,5-trichlorobenzene; 7, 1,2,4-trichlorobenzene; 8, 1,2,3-trichlorobenzene. (From ref. 38)

G

TABLE 8.8   Identification of sulphur compounds in garlic[86]

Observed ϕ^a	Expected type of compound	Observed relative retention time[b]	Assignment	Standard relative retention time[b]
0·4	Mercaptan	0·04	Methyl mercaptan	0·04
0·01	Monosulphide	0·06	Dimethyl sulphide	0·06
2·4	Disulphide	0·15	Dimethyl disulphide	0·16
0·09	Monosulphide	0·21	Diallyl sulphide	0·18
10·0	Unsaturated disulphide	0·39	Methyl allyl disulphide	0·46
300·0	Trisulphide	0·65	Dimethyl trisulphide	0·76
20·0	Unsaturated disulphide	1·00	Diallyl disulphide	1·00
200·0	Trisulphide	1·68	Methyl allyl trisulphide	1·63
320·0	Trisulphide	3·42	Diallyl trisulphide	3·42

[a] ϕ = Electron capture response/flame ionisation response.
[b] On a 10% Carbowax 20M column at 110°C.

The relative EC and FID responses of several low molecular weight halogenated substances have been studied,[88] and it has been found that the relative responses vary over seven orders of magnitude. While fluorinated alkanes have low EC responses, C_6F_6 gives a high response. In general, however, iodinated substances give the highest EC response, and there is a marked increase in relative response as the number of halogen atoms in the molecule increases. One iodine atom is roughly equivalent to three chlorine atoms.

Care should be taken when interpreting retention times measured from dual chromatograms for a parallel arrangement of detectors. Retention times for the same peak may appear to differ for two detectors unless the volumes of extra-column tubing and detectors are similar. Moreover, the apparent retention time of a multi-component peak may change if the detectors have very different sensitivities for the components.[89] For example, analysis of a mixture of two components A and B with a detector which is insensitive to component B will lead to a retention time which will be the true retention time of component A. If, however, the detector is sensitive to both A and B such that the electrical response is similar for each, then the observed retention time will lie between the true retention time for A and that for B. Such effects are a nuisance, but they may provide useful qualitative clues about the nature of the overlapped peaks.

REFERENCES

1. W. H. King, 'Piezoelectric sorption detector', *Analyt. Chem.*, **36**, 1735 (1964).
2. G. G. Guilbault, 'The use of mercury(II) bromide as coating in a piezoelectric crystal detector', *Analyt. Chim. Acta*, **39**, 260 (1967).
3. D. W. Turner, 'A robust but sensitive detector for gas–liquid chromatography', *Nature*, **181**, 1265 (1958).
4. J. D. Winefordner and D. Steinbrecker, 'A vapour detector based on changes in dielectric constant', *Analyt. Chem.*, **33**, 515 (1961).
5. J. D. Winefordner, H. P. Williams, and C. D. Miller, 'A high sensitivity detector for gas analysis', *Analyt. Chem.*, **37**, 161 (1965).
6. H. Purnell, *Gas Chromatography*, Wiley, New York (1962), pp. 302–304.
7. E. M. Bulewicz, 'Correlation between chemi-ionization in flames containing organic fuels and the heat of oxidation of the carbon in the fuel molecule', *Nature*, **211**, 961 (1966).
8. A. Karmen and L. Giuffrida, 'Enhancement of the response of the hydrogen flame ionization detector to compounds containing halogens and phosphorus', *Nature*, **201**, 1204 (1964).
9. L. Giuffrida, N. F. Ives, and D. C. Bostwick, 'Gas chromatography of pesticides: improvements in the use of special ionization detection systems', *J. Assoc. Offic. Agric. Chemists*, **49**, 8 (1966).
10. C. H. Hartmann, 'Aerograph phosphorus detector', *Aerograph Res. Notes*, 1 (1966) [*Analyt. Abs.*, **14**, 7244 (1967)].

11. N. F. Ives and L. Giuffrida, 'Investigation of thermionic detector response for the gas chromatography of P, N, As and Cl organic compounds', *J. Assoc. Offic. Agric. Chemists*, **50**, 1 (1967).

12. W. A. Aue, C. W. Gehrke, R. C. Tindle, D. L. Stalling, and C. D. Ruyle, 'Application of the alkali-flame detector to nitrogen-containing compounds', *J. Gas Chromatog.*, **5**, 381 (1967).

13. C. H. Hartman, 'Alkali flame detector for organic nitrogen compounds', *J. Chromatog. Sci.*, **7**, 163 (1969).

14. F. M. Page and D. E. Woolley, 'Mechanism of the determination of phosphorus with a flame ionization detector', *Analyt. Chem.*, **40**, 210 (1968).

15. M. Dressler and J. Janák, 'Detection of sulphur compounds with an alkali flame ionisation detector', *J. Chromatog. Sci.*, **7**, 451 (1969).

16. A. Karmen, 'Specific detection of halogens and phosphorus by flame ionization', *Analyt. Chem.*, **36**, 1416 (1964).

17. A. Karmen, 'Differential specificity in detecting phosphorus, nitrogen, and halogens with alkali flames', *J. Chromatog. Sci.*, **7**, 541 (1969).

18. J. Janák and V. Svojanovský, 'Working properties of a coupled flame ionization-sodium thermionic detector', in *Gas Chromatography 1966* (ed. A. B. Littlewood), Institute of Petroleum, London (1967), p. 166.

19. A. V. Nowak and H. V. Malmstadt, 'Selective gas-chromatographic detector utilizing emitted radiation from a sensitized flame', *Analyt. Chem.*, **40**, 1108 (1968).

20. J. E. Lovelock, 'A photoionization detector for gases and vapours', *Nature*, **188**, 401 (1960).

21. J. E. Lovelock and A. Zlatkis, 'A new approach to lead alkyl analysis. Gas phase electron absorption for selective detection', *Analyt. Chem.*, **33**, 1958 (1961).

22. D. C. Locke and C. E. Meloan, 'Study of the photoionization detector for gas chromatography', *Analyt. Chem.*, **37**, 389 (1965).

23. J. G. W. Price, D. C. Fenimore, P. G. Simmonds, and A. Zlatkis, 'Design and operation of a photoionization detector for gas chromatography', *Analyt. Chem.*, **40**, 541 (1968).

24. J. F. Roesler, 'Preliminary study of characteristics of photoionization detector for gas chromatography', *Analyt. Chem.*, **36**, 1900 (1964).

25. P. F. Deisler, K. W. McHenry, and R. H. Wilhelm, 'Rapid gas analyser using ionization by alpha particles', *Analyt. Chem.*, **27**, 1366 (1955).

26. F. W. Lampe, J. L. Franklin, and F. H. Field, 'Cross sections for ionization by electrons', *J. Amer. Chem. Soc.*, **79**, 6129 (1957).

27. D. J. Pompeo and J. W. Otvos, *U.S. Pat.* 2,641,710 (1953).

28. C. H. Deal, J. W. Otvos, V. N. Smith, and P. S. Zucco, 'A radiological detector for gas chromatography', *Analyt. Chem.*, **28**, 1958 (1956).

29. J. E. Lovelock, G. R. Shoemake, and A. Zlatkis, 'Sensitive ionization cross-section detector for gas chromatography', *Analyt. Chem.*, **35**, 460 (1963).

30. J. E. Lovelock, G. R. Shoemake, and A. Zlatkis, 'Improved ionization cross-section detectors', *Analyt. Chem.*, **36**, 1410 (1964).

31. K. Abel, 'An improved gas-tight micro-cross-section ionization detector', *Analyt. Chem.*, **36**, 954 (1964).

32. J. W. Otvos and D. P. Stevenson, 'Cross-sections of molecules for ionization by electrons', *J. Amer. Chem. Soc.*, **78**, 546 (1956).
33. H. Boer, 'A comparison of detection methods for gas chromatography including beta ray ionization', in *Vapour Phase Chromatography* (ed. D. H. Desty), Butterworths, London (1957), p. 169.
34. A. Zlatkis, G. R. Shoemake, and P. G. Simmonds, 'Determination of methyl anthranilate in grape juice', *J. Gas Chromatog.*, **5**, 19A (1967).
35. J. E. Lovelock, 'Affinity of organic compounds for free electrons with thermal energy: its possible significance in biology', *Nature*, **189**, 729 (1961).
36. D. C. Fenimore, A. Zlatkis, and W. E. Wentworth, 'Linearization of electron capture response by analog conversion', *Analyt. Chem.*, **40**, 1594 (1968).
37. J. E. Lovelock, D. C. Fenimore, and A. Zlatkis, 'Electron attachment spectroscopy', *J. Gas Chromatog.*, **5**, 392 (1967).
38. J. E. Lovelock, 'Ionization methods for the analysis of gases and vapours', *Analyt. Chem.*, **33**, 162 (1961).
39. V. Cantuti, G. P. Cartoni, A. Liberti, and A. G. Toni, 'Electron capture detector in gas chromatography. III. Comparative study of responses for several organic compounds', *J. Chromatog.*, **17**, 60 (1965).
40. P. Devaux and G. Guiochon, 'Variations of the response of the electron capture detector with carrier gas flow-rate', *J. Chromatog. Sci.*, **7**, 561 (1969).
41. E. S. Goodwin, R. Goulden, A. Richardson, and J. G. Reynolds, 'The analysis of crop extracts for traces of chlorinated pesticides by gas–liquid partition chromatography', *Chem. and Ind.*, 1220 (1960).
42. C. E. Cook, C. W. Stanley, and J. E. Barney, 'Correlation of structure of phosphate pesticides with response in electron affinity detectors', *Analyt. Chem.*, **36**, 2354 (1964).
43. M. J. de F. Maunder, H. Egan, and J. Roburn, 'Some practical aspects of the determination of chlorinated pesticides by electron-capture gas chromatography', *Analyst*, **89**, 175 (1964).
44. H. Egan, E. W. Hammond, and J. Thompson, 'The analysis of organophosphorus pesticide residues by gas chromatography', *Analyst*, **89**, 175 (1964).
45. W. L. Zielinski, L. Fishbein, and L. Martin, 'Relation of structure to sensitivity in the electron capture analysis of pesticides', *J. Gas Chromatog.*, **5**, 552 (1967).
46. R. A. Landowne and S. R. Lipsky, 'The electron capture spectrometry of haloacetates: a means of detecting ultramicro quantities of sterols', *Analyt. Chem.*, **35**, 532 (1963).
47. A. L. Henne, M. S. Newman, L. L. Quill, and R. A. Staniforth, 'Alkaline condensation of fluorinated esters with esters and ketones', *J. Amer. Chem. Soc.*, **69**, 1819 (1947).
48. R. E. Sievers, B. W. Ponder, M. L. Morris, and R. W. Moshier, 'Gas phase chromatography of metal chelates of acetylacetone, trifluoroacetylacetone and hexafluoroacetylacetone', *Inorg. Chem.*, **2**, 693 (1963).
49. R. W. Moshier and R. E. Sievers, *Gas Chromatography of Metal Chelates*, Pergamon Press, Oxford (1965).
50. W. D. Ross, 'Detection of metal chelates in gas–liquid chromatography by electron capture', *Analyt. Chem.*, **35**, 1596 (1963).

51. W. D. Ross and G. Wheeler, 'Quantitative determination of chromium(III) hexafluoroacetylacetonate by gas chromatography', *Analyt. Chem.*, **36**, 266 (1964).

52. J. L. Monkman and L. Dubois, 'The determination of halogenated hydrocarbons by gas chromatography and flame photometry', in *Gas Chromatography* (ed. H. J. Noebels, R. F. Wall, and N. Brenner), Academic Press, New York (1961), p. 333.

53. F. A. Blunther, R. C. Blinn, and D. E. Ott, 'Beilstein flame method of detection of organohalogen compounds emerging from a gas chromatograph', *Analyt. Chem.*, **34**, 302 (1962).

54. M. C. Bowman and M. Beroza, 'A copper-sensitized flame-photometric detector for gas chromatography of halogen compounds', *J. Chromatog. Sci.*, **7**, 484 (1968).

55. D. W. Grant, 'An emissivity detector for gas chromatography', in *Gas Chromatography 1958* (ed. D. H. Desty), Butterworths, London (1958), p. 153.

56. D. W. Grant and G. A. Vaughan, 'The use of gas–liquid chromatography in the determination of the distribution of aromatic compounds in coal tar naphthas', in *Vapour Phase Chromatography* (ed. D. H. Desty), Butterworths, London (1957), p. 413.

57. R. S. Juvet and R. P. Durbin, 'Flame photometric detection of metal chelates separated by gas chromatography', *J. Gas Chromatog.*, **1**, 14 (1963).

58. R. S. Juvet and R. P. Durbin, 'Characterisation of flame photometric detector for gas chromatography', *Analyt. Chem.*, **38**, 565 (1966).

59. F. M. Zado and R. S. Juvet, 'A new selective nonselective flame photometric detector for gas chromatography', *Analyt. Chem.*, **38**, 569 (1966).

60. R. W. Morrow, J. A. Dean, W. D. Shults, and M. A. Guerin, 'A silicon-specific detector based on interfacing a gas chromatograph and a flame emission or atomic absorption spectrometer', *J. Chromatog. Sci.*, **7**, 572 (1969).

61. R. S. Braman, 'Flame emission and dual flame emission-flame ionization detectors for gas chromatography', *Analyt. Chem.*, **38**, 734 (1966).

62. A. J. McCormack, S. C. Tong, and W. D. Cooke, 'Sensitive selective gas chromatography based on emission spectrometry of organic compounds', *Analyt. Chem.*, **37**, 1470 (1965).

63. S. S. Brody and J. E. Chaney, 'Flame photometric detector. Application of a specific detector for phosphorus and for sulfur compounds sensitive to subnanogram quantities', *J. Gas Chromatog.*, **4**, 42 (1966).

64. C. W. Stanley and J. I. Morrison, 'Identification of organophosphate pesticides by gas chromatography with the flame photometric detector', *J. Chromatog.*, **40**, 289 (1969).

65. M. C. Bowman and M. Beroza, 'Gas chromatographic detector for simultaneous sensing of phosphorus- and sulfur-containing compounds by flame photometry', *Analyt. Chem.*, **40**, 1448 (1968).

66. K. M. Aldous, R. M. Dagnall, S. J. Pratt, and T. S. West, 'Microwave excited electrodeless discharge tubes containing organo-sulfur and phosphorus compounds', *Analyt. Chem.*, **41**, 1851 (1969).

67. C. A. Bache and D. J. Lisk, 'Determination of organophosphorus insecticide residues using the emission spectrometric detector', *Analyt. Chem.*, **37**, 1477 (1965).

68. C. A. Bache and D. J. Lisk, 'Determination of iodinated herbicide residues and metabolites by gas chromatography using the emission spectrometric detector', *Analyt. Chem.*, **38**, 783 (1966).

69. C. A. Bache and D. J. Lisk, 'Low pressure emission spectrometric determination of part-per-billion residue levels of organophosphorus insecticides', *Analyt. Chem.*, **38**, 1757 (1966).

70. H. A. Moye, 'An improved microwave emission gas chromatography detector for pesticide residue analysis', *Analyt. Chem.*, **39**, 1441 (1967).

71. C. A. Bache and D. J. Lisk, 'Selective emission spectrometric determination of nanogram quantities of organic bromine, chlorine, iodine, phosphorus, and sulfur compounds in a helium plasma', *Analyt. Chem.*, **39**, 786 (1967).

72. C. A. Bache, L. E. St. John, and D. J. Lisk, 'Gas chromatography of insensitive pesticides as their halomethyldimethylsilyl derivatives', *Analyt. Chem.*, **40**, 1241 (1968).

73. R. S. Braman and A. Dynako, 'Direct current spectral emission-type detector', *Analyt. Chem.*, **40**, 95 (1968).

74. P. Hersch, 'Galvanic analysis', in *Advances in Analytical Chemistry and Instrumentation*, Vol. 3, Interscience, New York (1964), p. 183.

75. D. M. Coulson and L. A. Cavanagh, 'Automatic chloride analyser', *Analyt. Chem.*, **32**, 1245 (1960).

76. R. L. Martin and J. L. Grant, 'Determination of sulphur-compound distributions in petroleum samples by gas chromatography with a coulometric detector', *Analyt. Chem.*, **37**, 644 (1965).

77. E. M. Fredericks and G. A. Harlow, 'Determination of mercaptans in sour natural gases by gas liquid chromatography and microcoulometric titration', *Analyt. Chem.*, **36**, 263 (1964).

78. H. V. Drushel, 'Sulfur compound type distributions in petroleum using an in-line reactor or pyrolysis combined with gas chromatography and a microcoulometric sulfur detector', *Analyt. Chem.*, **41**, 569 (1969).

79. P. J. Klaas, 'Gas chromatographic determination of sulphur compounds in naphthas employing a selective detector', *Analyt. Chem.*, **33**, 1851 (1961).

80. R. L. Martin, 'Fast and sensitive method for determination of nitrogen. Selective nitrogen detector for gas chromatography', *Analyt. Chem.*, **38**, 1209 (1966).

81. H. L. Pease, 'Determination of terbacil residues using microcoulometric gas chromatography', *J. Agric. Food Chem.*, **16**, 54 (1968).

82. D. M. Coulson, 'Electrolytic conductivity detector for gas chromatography', (a) *J. Gas Chromatog.*, **3**, 134 (1965); (b) *The Detektor*, **1** (1) (1968), Micro Tek Instrument Corp., U.S.A.

83. D. M. Coulson, 'Selective detection of nitrogen compounds in electrolytic conductivity gas chromatography', *J. Gas Chromatog.*, **4**, 285 (1966).

84. A. Djkstra, C. C. M. Fabrie, G. Kateman, C. J. Lamboo, and J. A. L. Thisson, 'Recording conductometer for the determination of small amounts of carbon dioxide and its use in combination with the combustion technique in gas chromatography', *J. Gas Chromatog.*, **2**, 180 (1964).

85. H. J. Dawson, 'Detection of traces of polynuclear aromatics in hydrocarbons by gas chromatography', *Analyt. Chem.*, **36**, 1852 (1964).

86. D. M. Oaks, H. Hartman, and K. P. Dimick, 'Analysis of sulfur compounds with electron capture/hydrogen flame dual channel gas chromatography', *Analyt. Chem.*, **36**, 1560 (1964).
87. W. L. Zielinski, L. Fishbein, and R. O. Thomas, 'Simple effluent splitter for measurement of electron capture/flame ionization response ratios', *Analyt. Chem.*, **39**, 1674 (1967).
88. C. A. Clemons and A. P. Altshuller, 'Response of electron-capture detector to halogenated substances', *Analyt. Chem.*, **38**, 133 (1966).
89. C. L. Teitelbaum, 'Possible errors in interpreting results of two channel gas chromatography', *Analyt. Chem.*, **37**, 309 (1965).

CHAPTER NINE

PEAK-TRAPPING

Modern qualitative analysis relies heavily upon instrumental methods of identification, several of which can be applied directly to GC effluents. Direct monitoring is conceptually very attractive, but most analysis is still carried out by transferring the eluates from the chromatograph via some method of peak-trapping. Indeed, there are situations in which trapping is unavoidable:

(a) When using ancillary equipment with a scan rate which is slow in comparison with the rate of elution of the peak (cf. pp. 213—217 and 244—250). Then the peak can be trapped and a complete analysis carried out at leisure in a single run.

(b) In the collection of material from preparative-scale chromatographs. Several of the traps described in this chapter have been designed specifically for this purpose.

(c) In small laboratories where the limited availability of equipment at any one time may make it necessary to trap and store chromatographic fractions for analysis at some later date.

(d) When using several columns to identify samples by means of retention behaviour (Chapter Three), single peaks from one column are often trapped and re-injected on to another.

Peak-trapping has received constant attention since the beginning of GC, and yet still continues to present problems. A bewilderingly large number of different trap systems have been used, although the fundamental understanding of them has been small. As a result, the chromatographer who wishes to effect the apparently elementary act of converting a vapour into a liquid (or a solid) is faced with a formidable number of possibilities including packed and unpacked cold traps; solution and entrainment traps; total effluent and adsorption traps; potassium bromide, filter, and Volman traps; and electrostatic precipitators. In the next section of this chapter the physical background to trapping is considered, and an attempt is made to put the more important factors on a quantitative footing. This section is followed by an account of the types of trap in common use.

9.1 PHYSICAL BACKGROUND TO TRAPPING

The basis of any method of trapping is the transference of an eluate from the gas phase to some condensed phase. It follows that the very minimum

191

requirement for trapping to occur in any system is that the partial pressure of the eluate entering the trap must be greater than its equilibrium vapour pressure with the condensed phase at the trap temperature. Thus, provided that gaseous and condensed phases are equilibrated (unlikely in practice), the most efficient trapping system will simply be that in which the equilibrium vapour pressure of the eluate is a minimum.

The equilibrium vapour pressure of an eluate may be reduced by lowering the temperature, by dissolving the eluate in a solvent, by adsorbing it, or by forming a chemical complex. All these methods, except the last which is too specific for general use, have been used in trapping. However, by far the most common method of reducing the equilibrium vapour pressure of an eluate is to reduce the temperature of the effluent, either in a trap filled with material such as GC column packing or in a simple open-tube trap. Ideally, the performance of the latter is governed by the principles of liquid–vapour equilibria. In practice, however, kinetic effects also play an important part, i.e. the eluate does not stay in the trap for long enough to enable equilibrium to be reached. Instead, eluate tends to pass straight through the trap, either as a supersaturated vapour or as a fog composed of small liquid droplets. Both the ideal equilibrium behaviour and the effects of supersaturation are discussed below.

9.1a Ideal trapping efficiency

In most practical situations the gas pressure does not vary greatly throughout the trap. Thus, if the partial pressure of eluate is p at room temperature T, then in the absence of condensation the eluate will also have a partial pressure of p at the trap temperature T'. Whether or not a condensed phase can subsequently form depends on the relative values of p and P, the equilibrium vapour pressure of the eluate at the trap temperature. If $p > P$, trapping should take place and, assuming equilibration, the ideal percentage efficiency E will be given by

$$E = 100 \left(\frac{p - P}{p} \right) = 100 \left(1 - \frac{P}{p} \right) \qquad (9.1)$$

This equation also predicts negative efficiencies when $p < P$. These will only be physically realisable if the trap already contains some liquid eluate, since eluate will then tend to evaporate in an attempt to increase p towards the value P. If the trap is empty, however, then E must always be zero when $p < P$.

During the elution of a peak, the partial pressure of eluate varies with time, so that the overall efficiency, $\bar{E}$, after time t will be given by

$$\bar{E} = \frac{1}{t} \int_0^t E \, dt \qquad (9.2)$$

where the instantaneous values of E are given by equation (9.1). If the trap is empty at the start of trapping, then E must be zero for the forward edge of the peak, since here p must always be less than P. Similarly, if some liquid is deposited, E will become negative at the trailing edge of the peak where again $p < P$. It follows that it is impossible, even in principle, to trap a peak completely by cooling. Furthermore, there will be a certain sample size below which p will be smaller than P throughout the whole peak, so that $\bar{E}$ will be zero. It might be thought that this sample size is that at which p_{max}, the value of p at the peak maximum, is just equal to P. However, even if p_{max} is somewhat greater than P, the overall efficiency $\bar{E}$ may still be zero because of re-evaporation of the trapped eluate as p falls below P at the trailing edge of the peak. Thus, p must be considerably larger than P if any permanent trapping is to occur, unless care is taken to isolate the condensed material before the whole peak has emerged.

These points are illustrated in Figure 9.1(a) for the condition $p_{max} = 4 \cdot 8P$. The two traces represent the peak envelopes (partial pressure against time) at the inlet (top) and outlet (bottom) of the trap. Note that no trapping occurs until point A is reached, where $p = P$. Also, beyond B the amount of trapped eluate decreases, the partial pressure of eluate emerging from the trap being maintained at P until all the trapped material is re-evaporated. Hence the efficiency of the trap decreases if it remains connected to the column outlet beyond point B. By approximating the peak-shape to a triangle, the following expression for the efficiency if trapping were stopped at time B is obtained from equation (9.2):

$$\bar{E}_b = 100(1 - P/p_{max})^2 \qquad (9.3)$$

For $p_{max} = 4 \cdot 8P$ [the example of Figure 9.1(a)], this equation gives $\bar{E}_b \approx 62\%$. Similarly, $p_{max} = 20P$ leads to about 90% efficiency, while an efficiency of 99% would only be obtainable for $p_{max} = 200P$. It must be remembered that these equations refer to ideal conditions in which equilibration is achieved.

Equation (9.3) may be used to obtain $\bar{E}_b$ at various sample sizes. For this purpose p_{max} must be related to the sample size. Approximation of the gaussian-shaped elution peak, with standard deviation σ (expressed in units of time), to a triangle of height p_{max} and base 4σ, leads to

$$p_{max} = \frac{NRT'}{2F'\sigma} = \frac{NRT}{2F\sigma} \qquad (9.4)$$

where N is the sample size in moles, and F' and F are the carrier gas flow rates measured at trap temperature T' and room temperature T,

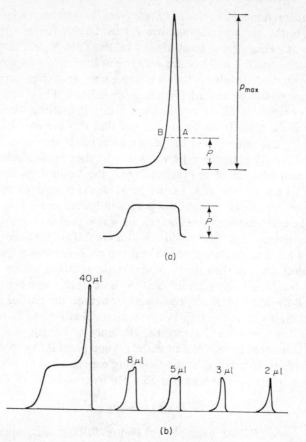

FIGURE 9.1 (a) Eluate and condensate in equilibrium: traces of the partial pressure of an eluate versus time at the inlet (top) and outlet (bottom) of an open-tube trap. Trapping commences beyond point A at which the eluate partial pressure is equal to the saturated vapour pressure, P, at the trap temperature. Beyond point B the amount of eluate held in the trap decreases as it bleeds off at partial pressure P (bottom).

FIGURE 9.1   (b) Experimental traces for n-pentane obtained at the outlet of an open-tube trap cooled in a solid carbon dioxide–trichloroethylene bath. Impressed on the 'ideal' trace given in Figure 9.1(a) (bottom) is a peak 'shadow' indicating that the effluent and condensate are not in equilibrium. This effect is probably due to aerosol formation. Flow rate at room temperature, 123·6 ml/min. Sample sizes refer to a 12·5% (v/v) solution of n-pentane in carbon tetrachloride. The height of the plateau corresponds to the vapour pressure of n-pentane at a temperature about 4·5° above that of solid carbon dioxide. This difference between the temperatures of the inside and outside of the trap is probably due to warming by the carrier gas. (From ref. 1) respectively. From equations (9.3) and (9.4) we obtain

$$\bar{E}_b = 100[1 - (2F\sigma P/NRT)]^2 \qquad (9.5)$$

Some values of $\bar{E}_b$ calculated from this equation for various sample sizes at the temperatures of ice and solid carbon dioxide are shown in Table 9.1 for $F = 30$ ml min^{-1} and $\sigma = 7$ sec.

TABLE 9.1   Theoretical values of $\bar{E}_b$ [see equation (9.5)] for several eluates at the temperatures of ice and solid carbon dioxide

Sample size (g)	At 0 °C			At −78 °C		
	Benzene	Pent-1-ene	n-Octane	Benzene	Pent-1-ene	n-Octane
10^{-2}	85%	9%	98%	∼100%	98%	∼100%
10^{-3}	4	∼0	79	∼100	86	∼100
10^{-4}	∼0	∼0	∼0	98	9	∼100
10^{-5}	∼0	∼0	∼0	85	∼0	99
10^{-6}	∼0	∼0	∼0	4	∼0	97

It is seen that the sample sizes required for any reasonable efficiency of trapping of relatively low molecular weight eluates are in accord with the known low efficiencies of cold traps in collecting trace amounts of material. Although the use of liquid nitrogen leads to a dramatic reduction of P, heat-transfer effects and aerosol formation become so acute that there is then little practical value in employing equation (9.5) to estimate an ideal trapping efficiency.

9.1b   Non-ideal trapping efficiency

The discussion so far has assumed a simple situation in which the liquid and vapour within the trap are assumed to be in equilibrium. In practice, however, this is unlikely to be the case, and the vapour will to various degrees become supersaturated, or form an aerosol. For this reason the concentration of eluate leaving the trap can be larger than that corresponding to P, so that the plateau AB of Figure 9.1(a) has impressed upon it a peak 'shadow', as shown in Figure 9.1(b) for the elution of n-pentane at the temperature of solid carbon dioxide.[1] It follows that the experimentally measured value of $\bar{E}_b$ will be below that calculated from equation (9.3). Thus, the ideal efficiencies given in Table 9.1 for various sample sizes are higher than those found in practice by some factor which will vary from case to case depending on the fraction of eluate in the peak 'shadow' of Figure 9.1(b). This will, in turn, be determined largely by the extent of aerosol formation.

Aerosol formation

According to equation (9.5), the efficiency of a cold trap should increase with sample size, N. This has been confirmed for the trapping of n-pentane at the temperature of solid carbon dioxide.[1] In contrast, however, in a quantitative study of the collection of steroids in traps at room temperature,[2] it was found that the efficiency *decreased* linearly with log sample size, as shown in Figure 9.2. These anomalous results can be rationalised in terms of aerosol formation.

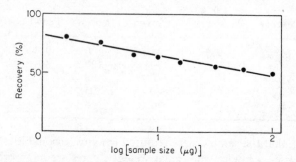

FIGURE 9.2   Semilogarithmic plot of the percentage recovery of microgram quantities of the steroid 17-β-oestradiol-4-[14C] diacetate in a melting-point tube (1 mm inside diameter × 10 cm) at room temperature. The observed decrease of efficiency with increasing sample size is contrary to simple theoretical considerations, and probably indicates considerable aerosol formation. Column: SE-30 on Chromosorb W (1·2% w/w) operated at 240°. (From ref. 2)

Aerosols (fogs) are formed from supersaturated vapours at rates which are very high at high degrees of supersaturation and/or in the presence of particles which can act as nuclei for droplets. In the absence of such nuclei, the rate-determining step for deposition is the formation of liquid nuclei of a critical size by the random aggregation of vapour molecules. In most chromatographic situations it is unlikely that the supersaturated vapour will be clean enough for the latter process to be of prime importance, and it is far more likely that the rate of growth of the aerosol will be determined by the presence of dust particles which are larger than, or equal in size to, the critical nucleus.

The critical radius is given by

$$r_{\text{crit}} = \frac{2sv_{\text{m}}}{kT \ln (p_{\text{s}}/P)} \tag{9.6}$$

where s is the bulk surface tension of the liquid, v_m is the volume occupied by a molecule in the liquid, p_s is the vapour pressure above the surface of the droplet, and P is the equilibrium vapour pressure over an infinite surface.[3] If an instantaneous trapping efficiency greater than 90% is to be possible, even in principle, $p_s/P > 10$ [see equation (9.1)]. Using this value together with reasonable bulk values of s and v_m, it follows that r_{crit} is about 10 Å at 300°K. Since v_m increases with molecular weight, so also will r_{crit}, but for most situations it will not exceed 100 Å except at very low values of p_s/P. Even though the extrapolation of bulk properties down to a microscopic level may not be strictly valid, it is evident that in most cases the critical nucleus is so small that the carrier-gas stream almost certainly contains dust particles which are of an equal or larger size, and upon which, therefore, droplets will form. Also it can be seen from equation (9.6) that, as p_s/P increases, r_{crit} decreases, so the number of such dust particles increases. Thus, for a particular substance, the number of supercritical nuclei available will increase with sample size.

We may put this on a more quantitative basis by employing the distribution function, $f(r)$, for the concentration of dust particles of radius r. The number of particles, m, of radius greater than r_{crit} in unit volume is then represented by

$$m = \int_{r_{crit}}^{\infty} f(r) \, dr$$

If it is assumed that the whole of the eluate forms an aerosol, and that all supercritical nuclei are involved, the average size of droplet in moles is given by

$$\bar{N} = \frac{p_s}{mRT} \tag{9.7}$$

The variation of this expression with p_s, and hence with the sample size, depends on the form of $f(r)$, and this will vary from case to case. The most important factor is the relative value of r_{crit} and r_{max}, the most probable particle radius. This is illustrated schematically in Figure 9.3 for a hypothetical distribution; the shaded areas represent the values of m, i.e. the number of supercritical nuclei. If $r_{crit} \ll r_{max}$, m will remain almost constant if r_{crit} is reduced as a result of an increased sample size [Figure 9.3(a)]. Thus, according to equation (9.7), the average droplet size $\bar{N}$ will increase with sample size. On the other hand, if $r_{crit} \gg r_{max}$, a slight reduction in r_{crit} as a result of increased sample size will lead to a significant increase in the number of available supercritical nuclei, m [Figure 9.3(b)]. Under these conditions, therefore, an increase in p_s, i.e. an increased sample size, will result in a fall in p_s/m, so that, according to equation (9.7), the average droplet size $\bar{N}$ will decrease. Since the average size of droplet probably determines the efficiency of retention of droplets within a given trap, greater trapping efficiencies being obtained for higher values of $\bar{N}$, it is seen that the effect of aerosol formation upon overall trap efficiency may either increase or decrease with sample size, depending on circumstances. The results of Figure 9.2, where lower trapping efficiencies were found for larger sample sizes, can thus be rationalised if $r_{crit} \gg r_{max}$ for the conditions employed. Experimental data to test these speculations do not seem to be available.

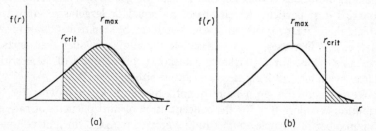

FIGURE 9.3   Hypothetical size distribution of dust particles in carrier gas for the cases (a) $r_{crit} \ll r_{max}$ and (b) $r_{crit} \gg r_{max}$. In case (a) the concentration of supercritical nuclei, m (shaded area), would not increase greatly if r_{crit} were reduced. In case (b) m would be very sensitive to the value of r_{crit}.

9.2 TYPES OF TRAP

The ideal trap is cheap, easy to use, and has a high efficiency. The very simplest traps are likely to be of greatest general use, and should be tried first. The simple, unpacked U-trap, although limited by sample size considerations, as shown above, may be useful for trapping large amounts of material. In other cases it may be necessary to resort to more complicated arrangements, and indeed many of the traps described below were designed to overcome deficiencies in simpler methods, particularly in relation to aerosol formation.

Before adopting a particular type of trap, its efficiency should always be determined with standard samples. The obvious way to do this is to inject a fixed amount of material and to determine the resultant peak areas with, and then without, the trap. If a non-destructive detector such as a katharometer is used, this operation may be carried out in a single step by interposing the trap between two detectors connected in series after the column, and measuring the response of both detectors to a given sample. A dual-channel katharometer may be used in a similar manner.

9.2a   Temperature-gradient traps

The cold-trapping of eluates with boiling points greater than about 150 °C is almost inevitably accompanied by the formation of aerosols which are then swept straight through the trap. One method of minimising fog formation is to pass the effluent through a gradual change of temperature instead of the more usual, abrupt change. Droplet formation is then made to occur at lower values of the supersaturation ratio, p_s/P, and hence requires a larger critical radius [equation (9.6)]. If the carrier-gas stream contains sufficiently large dust particles, this situation should lead to increased efficiency, since on average larger droplets will be formed.

Table 9.2 shows results obtained for the trapping of n-pentane in a simple U-trap cooled in liquid nitrogen. It can be seen from these data that the efficiency of such an arrangement is virtually independent of the depth of immersion of the trap in the refrigerant but depends strongly on the depth of the air space in the flask; the smaller this is the less efficient the trap becomes. This result emphasizes the importance of temperature-gradient effects in trapping.

TABLE 9.2   Variations in the efficiency of trapping of n-pentane in a U-trap cooled in liquid nitrogen[1] (sample size 0·6 μl; column temp. 42 °C; helium flow 125 ml/min)

(a) $l_2 = 51$ mm[a] l_1 (mm)	E (%)	(b) $l_1 = 38$ mm[a] l_2 (mm)	E (%)
3	68·0	100	96·4
18	69·7	95	84·5
69	71·0	74	87·5
100	75·4	58	75·2
		33	63·0

[a] l_1 = Depth to which trap is immersed in the liquid nitrogen; l_2 = distance of liquid-nitrogen surface below rim of Dewar vessel.

Figure 9.4(a) shows a temperature-gradient trap consisting of a piece of Teflon tubing which passes into a collecting vessel with a small nipple at its lower end. Condensed material is centrifuged into this nipple and may then be removed with a syringe. The trap has been found suitable for milligram quantities of oxygenated terpenes.[4] Figure 9.4(b) shows another simple type of trap suitable for peak loads in the 5—100 mg range, in which the temperature gradient is set up along a plug of steel wool. By careful positioning of this plug, aerosol formation with its attendant loss of efficiency can be minimised. However, in practice the condensation of an eluate in any trap is never complete, and at least part of it will exist as an aerosol. If this aerosol is now re-evaporated and passed through a second temperature gradient, a further amount of eluate will be condensed, thereby improving the overall efficiency of the trap. This is the principle of the

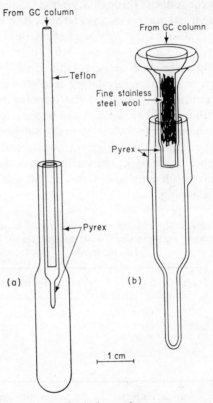

FIGURE 9.4   Two simple designs of temperature-gradient traps. After collection, the condensed material is centrifuged into the nipple of the collecting vessel. Proper positioning of the stainless steel wool plug in (b) is important to prevent aerosol formation. Trap (a) has been found suitable for 0·5—5 mg samples of oxygenated terpenes, while (b) is useful at the higher range 5—100 mg. (From ref. 4)

multiple temperature-gradient trap, which has, for example, been used for the trapping of fatty-acid esters.[5] Figure 9.5 shows a simple design of this type of trap which has been found to give quantitative recovery of milligram quantities of n-pentane.[1]

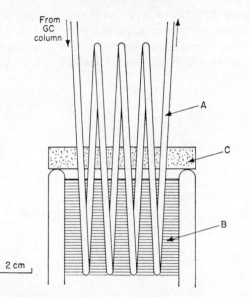

From
GC
column

A

C

B

2 cm

FIGURE 9.5   Multiple temperature-gradient trap in which the
effluent passes repeatedly between the refrigerant B and room
temperature in the glass spiral A. The polystyrene slab C helps
to maintain a temperature gradient

9.2b   Volman trap

Turbulent flow within a trap facilitates the collision of eluate droplets with
the wall, thereby inducing their retention. Also, heat transfer between
the trap wall and the effluent is improved. Both of these factors are
probably of importance in explaining the effectiveness of a type of trap
first described by Volman.[6] The trap consists of a double-walled vessel
within which turbulence is created by maintaining its walls at different
temperatures. Figure 9.6 shows a simple design of the trap. The Volman
trap is particularly useful for recovering fractions from preparative scale
columns. For example, gram quantities of ethyl caproate and limonene
have been recovered, with an efficiency of greater than 95%, using the
apparatus of Figure 9.6 with the inner wall at the temperature of solid
carbon dioxide and the outer wall at 150—200 °C.[7]

9.2c   Electrostatic precipitators

Since aerosols contain large numbers of charged droplets, it is evident that
trapping efficiencies should be improved by passing the cooled effluent
through a large electrostatic field. Indeed, it is known that electrostatic

effects influence the deposition of aerosols even at gas velocities as high as 7 m/sec. Electrostatic precipitation has been used in GC, particularly in connection with preparative-scale work. For example, the efficiency of trapping gram quantities of various high-boiling eluates was increased

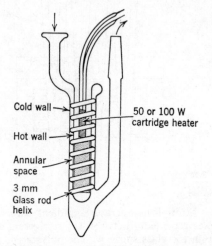

FIGURE 9.6   A Volman trap in which turbulence created between the hot and cold walls causes aerosol droplets to impact and condense on the walls. (From ref. 7)

from 10 to 90% by subjecting the cooled effluent to a silent discharge from an automobile ignition coil.[8] Samples ranging from a few milligrams to a few grams have been trapped with efficiencies up to 98%. Any source of high voltage in the region 10—50 kV can be used, the only requirement being that the effluent is not subjected to a discharge as this will undoubtedly cause some decomposition of the eluate.

9.2d   Packed traps

Aerosol formation in a simple U-trap can be greatly reduced by filling the trap with a packing material of high surface/volume ratio, such as glass wool or a GC support material. This is particularly useful for volatile eluates, since full advantage can then be taken of the low vapour pressure at liquid-nitrogen temperatures. It is important to ensure that adsorption on the packing is reversible, and deactivated support material is obviously most suitable. An essentially similar trap comprising a fritted filter has been used to recover samples of from 0·2 to 2 g with 95—97% efficiency.[9]

GC support coated with typical liquid phases has also been used for trapping. Even without refrigeration this allows considerable reduction of vapour pressure as a result of partition of the eluate between gas and liquid. The method is equivalent to isolating a peak in a portion of the column. This approach is particularly useful for trapping peaks prior to re-injection when attempting identification from retention behaviour, as discussed in Chapter Three. The volume of the trap must be suited to the peak being trapped. If no refrigeration is used, then it must be remembered that although substances with the shortest retention times have smallest peak widths expressed in terms of volume of carrier gas, they occupy the *greatest* length of column, since they are least soluble in the liquid phase.

Table 9.3 lists recoveries found after trapping and re-injection of a number of compounds of widely different volatilities. The trap used was a 4-inch packed column cooled to the temperature of solid carbon dioxide during trapping. After warming to about 200 °C, re-injection was effected

TABLE 9.3.   Recoveries of small amounts of eluates after trapping and re-injection[10]

Compound	Sample size (μg)	Recovery (%)
Heptan-1-ol	10·0	75
trans-Dodec-7-en-1-ol	1·0	76
Acetophenone	1·0	98
Octan-2-one	0·1	58
Nonanal	1·0	98
Methyl formate	5·0	64
Methyl propionate	0·1	154
Methyl butyrate	2·0	108
Methyl hexanoate	0·01	77
Methyl octanoate	0·01	86
Methyl myristate	1·0	92
Pentane	3·0	96
Octane	1·0	107

by flushing the trap with carrier gas from a syringe.[10] The percentage recoveries (found from the recorder trace of the second chromatogram) are good even below the microgram level. The trap has also been used for transferring as little as 10^{-7} g to a mass spectrometer. Miniature packed columns made of capillary tubes have also been used successfully for quantities of about 10^{-11} mole.[11]

A useful arrangement for trapping peaks prior to re-injection is shown in Figure 9.7. It incorporates three 3-way valves, and with these in configuration A the trap is isolated and the effluent bypassed to the detector.

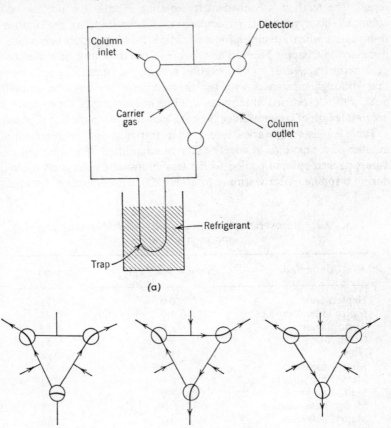

FIGURE 9.7 Device for trapping and re-injecting eluates based upon a triangular arrangement of three-way valves. (a) General flow diagram. (b) A, trap isolated; B, trapping effluent; C, re-injecting trapped material. (From ref. 12)

By watching the detector output, the valves are switched to arrangement B when it is required to trap an eluate. In this mode, the eluate passes through the cold-trap, where most of it is retained; the remainder passes on to the detector. When the required peak has been trapped, the valves are returned to configuration A and the trap is warmed. Meanwhile the columns are switched. Finally, conversion to the arrangement C causes the trapped peak to be re-injected on to the new column, with the trap acting as a sample loop.

The duration of trapping is chosen by referring to a complete chromatogram of the mixture, and varies according to the width of the peak and individual requirements. Sometimes, for example, only the leading edge of a complex peak will be wanted.

If the trap offers any significant resistance to flow, the retention volume of a material injected through the trap may differ from that injected in any other way. Care should always be taken, therefore, to compare retention data obtained with the same injection system.

9.2e Total-effluent traps

Perhaps the most obvious way to trap a material is to pass the whole of the effluent into an evacuated vessel. Alternatively, the effluent may be passed into a modified soap-bubble flowmeter with a small rubber or Teflon piston inserted in the main tube. At the appropriate moment the effluent is switched into the flowmeter tube and the piston is allowed to advance until the peak is trapped. Flow-through gas cells for infrared analysis, which also come under this heading, are discussed in Chapter Ten.

9.2f Entrainment techniques

With a condensable carrier gas such as argon or carbon dioxide, effluent trapping can be effected by condensing the total effluent, thereby entraining the small amounts of eluate in much larger quantities of carrier gas. The method is generally used in conjunction with argon carrier gas, and high efficiencies are attainable. Recoveries of around 95% for steroid samples up to 200 μg have been reported,[2] while 0·6—26 mg samples of various organic materials have been trapped with an efficiency of $100 \pm 3\%$.[14]

In order to minimise disturbances to the detector, the effluent must be trapped at such a rate that the pressure in the trap, which can be gauged by watching a soap-bubble flowmeter attached to its exit, does not fall far below atmospheric. When the required eluate has been trapped, the trap is disconnected, the condensed carrier gas is allowed to evaporate slowly, and the eluate is then recovered. A convenient method of gently warming the condensate is to centrifuge it. Carbon dioxide has some advantages over argon as a carrier gas. It is cheaper and condenses more readily, and since carbon dioxide is removed by sublimation, any possible loss of sample by ebullition is eliminated. On the other hand, argon can be removed at lower temperatures than carbon dioxide, and may therefore be more suitable for more volatile eluates.

In a related method the eluate is entrained in some easily condensable solvent mixed as a vapour with the effluent at the column outlet. In the original experiments, Jones and Ritchie, using water as solvent, were able to recover milligram amounts of materials such as toluidine and xylidine from preparative-scale columns with an efficiency of better than 95%.[15] A stream of solvent vapour was passed through a preheater to heat it to the column temperature, and was then mixed with the effluent before being passed through a condenser. A similar unit employing carbon tetrachloride as the solvent, and designed specifically for use with infrared micro-cells, is now available commercially.[16]

When using carbon dioxide carrier gas, absorption in alkali is an efficient method of collection (ref. 13; see p. 139). Gases can be recovered directly, and liquids and solids by solvent extraction.

9.2g   Trapping techniques for spectroscopy

Although the direct spectroscopic monitoring of chromatographic effluents is being used increasingly as a means of identifying eluates, the equipment required is not cheap, and it is not as generally available as more conventional equipment. There is still, therefore, a large need for methods of transferring eluates from traps into spectroscopic equipment. A review of methods of trapping fractions for infrared analysis has been published.[17]

In an attractive technique for trapping very small samples ($\sim 10^{-11}$ mole) prior to introduction into a mass spectrometer, the GC column is effectively extended by a short, thin-walled capillary filled with a convenient column packing.[11] Any given part of the effluent may then be trapped in this mini-column which is easily sealed and stored until it can be conveniently analysed. Empty capillary tubes suitably cooled can also be used.[18,19]

Several standard methods of preparing samples for infrared analysis have been adapted for use with GC. For example, eluates may be collected in a suitable solvent, either by bubbling or by using an entrainment technique (see above), and the resulting solution syringed into an infrared micro-cell. Alternatively, if the eluate is sufficiently involatile, the solution may be applied to a potassium bromide disc and the solvent evaporated. Potassium bromide discs have also been used in other ways. Eluates which condense readily at room temperature can be collected in small tubes packed with potassium bromide powder (10—100 mg). The powder is then pressed into a disc in the usual fashion. The method is particularly suitable for eluates of low volatility, and its extreme simplicity does much to commend it (see p. 215).

Several designs of infrared micro-cell are available commercially. They are made from a material such as sodium chloride or silver chloride which is transparent in the infrared, and have internal volumes which range from

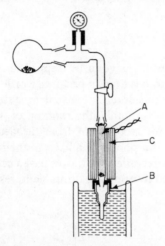

FIGURE 9.8 Apparatus for distilling eluates trapped on the mini-column A into the spectroscopic cell B. The apparatus is filled with argon. Eluates volatilised by the heater C are flushed into the liquid-nitrogen cooled cell (in this case a silver chloride infrared cell) while the argon freezes. (From ref. 20)

a few microlitres down to 0·2 μl. Corresponding pathlengths are from 0·1 to 0·01 mm. The cells are usually filled by centrifuging trapped material from a larger 'reception' volume into the micro-volume. In favourable cases the cells themselves may be used as traps. Alternatively, trapped material may be transferred to the micro-cells from some other trap. Fowlis and Welti have described a method for doing this when trapping is carried out on small chromatographic columns placed after the main column.[20] These columns are fitted into the apparatus shown in Figure 9.8, which includes a bulb of argon. The eluate is distilled from the trap into the micro-cell by heating the trap and immersing the micro-cell in a flask of liquid nitrogen. The eluate is flushed from the trap by freezing the argon. The argon is then allowed to evaporate from the cell which is then spun in a centrifuge. The cells used had pathlengths of 0·025 and 0·01 mm, and held 1 and 0·2 μl of liquid respectively. The method was applied to samples as small as 50 μg with trapping efficiencies around 90%. Similar results were obtained with NMR micro-cells made from melting-point tubes.

An ingenious means of interfacing a gas chromatograph with a spectrometer involves a flowing liquid interface. Eluates are trapped in a liquid and the resulting solution is passed through a cell for examination. The device has been used in connection with spectrophotofluorometry[21] and colorimetry.[22]

9.2h Automatic traps

If a large number of peaks have to be trapped, or repeated analyses performed, and a simple trap is used, it becomes worthwhile to consider automating the trapping procedure. A convenient system involves a number of traps mounted on a turntable, the rotation of which is triggered by a signal from the detector recorder or from the detector itself. One such arrangement consists of a photocell affixed to the recorder which triggers the turntable when the recorder pen moves past a certain pre-set point.[23] A limit switch is arranged to bring the turntable to rest when the new trap is

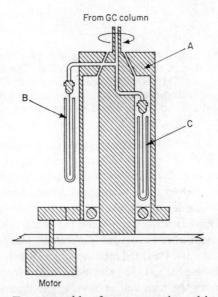

FIGURE 9.9   Trap assembly of an automatic multi-trap system. The turntable A brings the GC effluent in turn into each of twelve traps such as B. Pressure surges are avoided by means of the by-pass trap C.

in position. In addition to positioning a fresh trap for each peak, it is necessary to divert the effluent into this trap. A simple device which enables this to be done is shown in Figure 9.9. It includes a by-pass cell to prevent the build-up of an excessive back pressure in the column during switching.

REFERENCES

1. B. C. Shurlock, unpublished work.
2. S. C. Brooks and V. C. Godefroi, 'Quantitative collection of micro amounts of steroids from gas–liquid chromatography', *Analyt. Biochem.*, **7**, 135 (1964).
3. B. J. Mason, 'Nucleation of water aerosols', *Discuss. Faraday Soc.*, **30**, 20 (1960).
4. R. Teranishi, R. A. Flath, T. R. Mon, and K. L. Stevens, 'Collection of gas chromatographically purified samples', *J. Gas Chromatog.*, **3**, 206 (1965).
5. H. Schlenk and D. M. Sand, 'Collection of gas–liquid chromatography fractions by gradient cooling', *Analyt. Chem.*, **34**, 1676 (1962).
6. D. H. Volman, 'The mercury photosensitized reaction between hydrogen and oxygen', *J. Chem. Phys.*, **14**, 707 (1946).
7. R. Teranishi, J. W. Corse, J. C. Day, and W. G. Jennings, 'Volman collector for gas chromatography', *J. Chromatog.*, **9**, 244 (1962).
8. P. Kratz, M. Jacobs, and B. M. Mitzner, 'A smoke-eliminating device for a vapour-phase chromatographic fraction collector', *Analyst*, **84**, 671 (1959).
9. E. C. Schluter, 'Preparatory trapping of gas chromatographic effluents with a fritted filter', *Analyt. Chem.*, **41**, 1360 (1969).
10. B. A. Bierl, M. Beroza, and J. M. Ruth, 'Collection and transfer device for gas chromatographic fractions', *J. Gas Chromatog.*, **6**, 286 (1968).
11. J. W. Amy, E. M. Chait, W. E. Baitinger, and F. W. McLafferty, 'A general technique for collecting chromatographic fractions for introduction into the mass spectrometer', *Analyt. Chem.*, **37**, 1265 (1965).
12. D. A. Leathard and B. C. Shurlock, 'Gas chromatographic identification', in *Progress in Gas Chromatography* (ed. J. H. Purnell), Interscience, New York (1968), p. 1.
13. C. A. Bache and D. J. Lisk, 'A total collection gas chromatographic system—absorption of carbon dioxide carrier gas in alkali', *J. Chromatog. Sci.*, **7**, 296 (1969).
14. P. A. T. Swoboda, 'Total trapping of chromatographic effluents in argon carrier gas', *Nature*, **199**, 31 (1963).
15. J. H. Jones and C. D. Ritchie, 'A new procedure for the collection of fractions in gas chromatography', *J. Assoc. Offic. Agric. Chemists*, **41**, 753 (1958).
16. 'Carbon tetrachloride solution fraction collector for GC-IR', Bulletin 6803, Carle Instruments, Inc., California (1967).
17. J. G. Grasselli and M. K. Snavely, 'Methods for recovering gas chromatography and thin-layer chromatography fractions for infrared spectroscopy', *Progress in Infrared Spectroscopy*, **3**, 55 (1967).
18. K. R. Burson and C. T. Kenner, 'A simple trap for collecting gas chromatographic fractions for mass spectrometer analysis', *J. Chromatog. Sci.*, **7**, 63 (1969).

19. F. Armitage, 'Some factors affecting the efficiency of collection of gas chromatographic fractions in melting point tubes', *J. Chromatog. Sci.*, **7**, 190 (1969).
20. I. A. Fowlis and D. Welti, 'The collection of fractions separated by gas–liquid chromatography', *Analyst*, **92**, 639 (1967).
21. M. C. Bowman and M. Beroza, 'Apparatus combining gas chromatography with spectrophotofluorometry by means of a floating liquid interface', *Analyt. Chem.*, **40**, 538 (1968).
22. T. E. Healy and P. Urone, 'Gas chromatography of oxidants using a flowing liquid colorimetric detector', *Analyt. Chem.*, **41**, 1777 (1969).
23. G. Kronmueller, 'Automatic fraction collector for gas chromatography', in *Gas Chromatography*, Instr. Soc. Amer. Symposium, June 1959 (ed. H. J. Noebels, R. F. Wall, and N. Brenner), Academic Press, New York (1961), p. 199.

CHAPTER TEN

INFRARED AND OTHER SPECTROSCOPIC TECHNIQUES OF IDENTIFICATION

The principal reason for the extensive interest in direct coupling of GC with mass spectrometry (Chapter Eleven) is that both techniques are inherently suitable for sub-microgram quantities of material. In contrast, infrared and other spectroscopic techniques demand samples of about a milligram, and cannot therefore be so readily applied to GC. Recently, however, progress has been made in obtaining reasonable infrared spectra with much smaller samples by using miniature cells. Also, modifications to IR optical systems have resulted in much faster scanning, and hence facilitated on-line analysis of GC peaks.

10.1 INFRARED SPECTROSCOPY

IR spectroscopy is a commonly used general method for identifying organic compounds. Although the technique is largely empirical, the IR spectrum of a *pure* compound is extremely characteristic. This is not the place to consider the spectroscopic aspects of identification by IR, since there are already many excellent texts (e.g. refs. 1—4) which cover this subject. Instead, we shall consider the implications of mating GC and IR.[5] Since samples of a high degree of purity are needed for IR analysis, there is an obvious attraction in coupling IR with GC; the major drawback is the limited sensitivity of IR analysis.

10.1a Size of sample

The fraction α of incident light transmitted by a substance at a particular wavelength falls exponentially with increasing path-length l and concentration c according to the Beer–Lambert law:

$$\alpha = 10^{-\varepsilon cl}$$

where ε, a function of wavelength, is known as the molar absorptivity. Conventionally, c is expressed in moles per litre and l in cm, so that ε has units of $l\,mole^{-1}\,cm^{-1}$. Typical values of ε for IR absorption range from about $1\,l\,mole^{-1}\,cm^{-1}$ for a 'weak' band to $100\,l\,mole^{-1}\,cm^{-1}$ for a 'strong' band. If α (known as the transmittance) is very large, i.e. close to unity, then $(1-\alpha)$ is approximately equal to $2\cdot3\varepsilon cl$. Hence, when there are N

211

moles of sample in an IR beam of area a cm², $cl = 10^3 N/a$, so that for a *weak* absorption we may write,

$$(1 - \alpha) \approx \frac{2 \cdot 3 \times 10^3 \, \varepsilon N}{a} \tag{10.1}$$

IR instruments commonly have a response that is linear in α from unity (100% transmittance) to zero (total absorption). Thus, the deflection of the pen is a direct measure of $(1 - \alpha)$. Note that equation (10.1) is independent of the path-length, the maximum sensitivity for a given sample being obtained when it is *completely contained* within the beam.

Values of $(1 - \alpha)$ down to about 0·01 (i.e. 99% transmittance) can be detected and recognised as a band, which for a weak band ($\varepsilon = 1$) would correspond to the situation $N/a \approx 4 \times 10^{-6}$ moles/cm². A practical lower limit for a with conventional equipment is about 0·1 cm², so that the detection limit for a weak band is at about 0·4 μmole, i.e. typically 10^{-4} g. As discussed in Chapter Nine, it is possible to cold-trap this amount of eluate with reasonable efficiency. Hence, it is not surprising that, from the earliest days of GC, IR spectroscopy and cold-trapping have been used for the analysis of eluates at these levels. For a strong band with ϵ in excess of 100 the detection limit can be lowered to about 10^{-6} g. However, this is not always very useful since the fine-structure of IR spectra, composed of weak bands, is an important aspect of IR identification.

Reference to equation (10.1) shows that the obvious way to obtain maximum sensitivity for a given instrument is to ensure that both sample and beam are concentrated into the smallest possible value of a without, of course, decomposing the sample. a can be minimised by the use of a beam condenser. Figure 10.1(a) shows the IR spectrum of 1 μg of the pesticide Sevin in a 2 mm diameter disc of 4 mg potassium bromide using a beam condenser. Figure 10.1(b) shows the same spectrum run with 'scale expansion' of $\times 5$. Only a modest scale expansion is usually worthwhile, since most instruments show significant noise even on the normal scale. Another way in which a can be minimised is to place the sample as near as possible to an image of the source slit. If this procedure is adopted, great care must be taken to align the sample cell at the focus, both angularly and laterally.[7] Figure 10.2(a) shows spectra of benzyl acetate in carbon tetrachloride obtained in this way for sample sizes between 1 and 20 μg. The cylindrical micro-cells used had volumes of 1 or 2 μl. The spectrum of a large sample (2·3 mg) is shown for comparison in Figure 10.2(b). It should be noted, however, that not all commercial spectrometers have their beam focused in the sample compartment, so that an optical modification may be needed.

There is intense commercial interest in the field of GC–IR, and doubtless instruments suitable for very small samples will be forthcoming. At present, however, it is only practicable to use routine GC–IR in the analysis of samples which are larger than 10^{-5} g and preferably around 10^{-4} g.

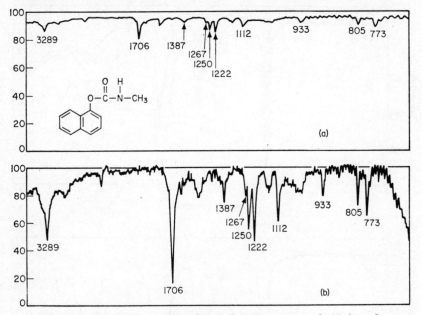

FIGURE 10.1   Micro KBr disc infrared spectra of 10^{-6} g of Sevin. (a) Without scale expansion, (b) × 5 scale expansion. (From ref. 6)

10.1b   Cells for trapped samples

When using conventional IR equipment, a choice must be made between trapping the eluate in a gas cell or attempting to condense it. Despite the difficulties inherent in the condensation procedure, this approach seems to be most common for all but the lowest-boiling eluates. Among the reasons for this is that liquid-phase spectra are better documented than those in the gas phase, although differences are usually slight for all but polar and hydrogen-bonding molecules. The short path-length of liquid samples is also an advantage, since it allows the whole sample to be held close to the focus of the sample beam, as discussed above.

Condensed samples

The prime difficulty associated with trapping, as discussed in Chapter Nine, is the formation of aerosols. If reliance is placed merely on cooling, substantial loss of sample is likely, and a 50% recovery must often be regarded as very satisfactory. Other serious losses of sample occur in the transference of eluate from the trap to the IR cell. For small samples it is

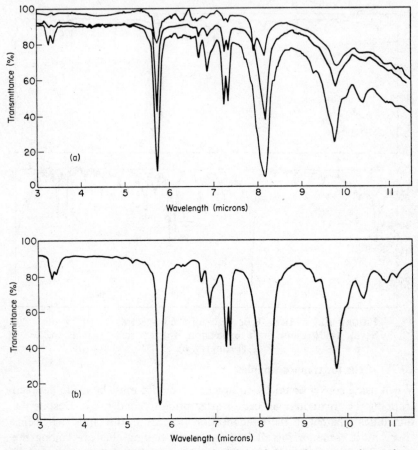

FIGURE 10.2   Infrared spectra of solutions of benzyl acetate in carbon tetrachloride. (a) Microcell spectra from 1, 5, and 20 μg of benzyl acetate. (b) Comparison spectrum from 2·3 mg of benzyl acetate. (From ref. 7)

therefore advantageous to use the trap as the cell. Suitable micro-cells made of a material transparent in the IR, such as silver chloride, are available commercially[8] and have been used for the simple cold-trapping at $-70\,^{\circ}$C of sample sizes of the order of 10^{-2} g with about 60% efficiency for acetone and toluene.[9] Very low efficiencies of recovery for volatile substances are obtained with samples smaller than 10^{-3} g unless a solvent such as carbon tetrachloride is present in the cell. Co-condensation of eluate and solvent vapour can also be used (p. 206).

IR micro-cells typically have a diameter (equal to path-length) in the range 0·01—0·10 mm with a constant ratio of volume to path-length of 20 μl/mm. It appears that sensitivity greater by a factor of 2 or 3 is obtainable by the use of a trap which serves also as a multiple internal-reflection cell.[10] This can have an effective path-length of 0·03 mm for an actual film thickness of 10^{-3} mm over an area of about 2 cm², corresponding to a volume of about 0·2 μl ($\sim$0·2 mg). However, it is not practicable to use this device below about 0 °C, whereas other micro-cells can be cooled to liquid-nitrogen temperatures without difficulties arising.

The potassium bromide micro-disc technique (see p. 206) is particularly attractive for small sample sizes of fairly involatile material. Samples can be trapped on potassium bromide mini-columns attached to the exit port of the gas chromatograph, or trapped in some other way and applied to the potassium bromide in a volatile solvent.[11] Alternatively, the column effluent can be applied directly to potassium bromide powder cooled by contact with refrigerated metal. Aerosol formation can be minimised by actually placing the gas chromatograph exit port in the powder. To improve the collection efficiency of the more volatile eluates, the potassium bromide can be doped with a little stationary phase, such as petroleum jelly, although this blocks parts of the IR spectrum. Simple micro-pellet dies are available for the production of clear discs from the powder as small as 2 or 3 mm diameter. Normal sized pellet dies can also be adapted to produce good micro-pellets with diameters as small as 1·5 mm.[12] Given a reasonable trapping efficiency, it is possible to obtain spectra from potassium bromide discs that permit identification below the 10^{-5} g level[6,11,12] (see Figure 10.1).

Gas cells

The main attraction of a gas cell, particularly for volatile samples, is the possibility of 100% trapping efficiency. At the same time gas-phase trapping makes possible, at least in principle, direct on-line coupling of GC and IR, as discussed in Section 10.1c below. The main drawback of gas-phase trapping is the need for a large cell. Further, since it is desirable to arrange to have all the sample in the IR beam [see equation (10.1)], a long narrow cell is required. For a beam of uniform area a, the length l required to obtain a band for a cell of volume V can be written as

$$l = V/a = (N/a)_{\mathrm{d}} \times (V/N) \qquad (10.2)$$

H

where N is the number of moles of sample and $(N/a)_d$ is the number of moles per cm^2 required for detection of the band, which was shown above (p. 212) to be about 4×10^{-6} moles/cm^2 for a weak band.

The width of a gaussian GC peak measured as the distance cut off on the time axis by tangents to the inflection points is given by

$$W = 4V_R/\sqrt{n} \qquad (10.3)$$

where V_R is the retention volume and n is the number of theoretical plates. Table 10.1 shows values of peak-widths, W, expressed as ml of carrier gas at the outlet pressure, for various retention times on a number of typical

TABLE 10.1   Typical widths of gas chromatographic peaks expressed in terms of volume of carrier gas containing them [equation (10.3)]

Column (theor. plates)	Flow rate (ml/min)	Retention time from injection	Approximate peak-width (ml)
1000	50	20 sec	2
		2 min	13
		10 min	63
		1 hr	380
10,000	20	5 min	4
		20 min	16
		1 hr	48
		3 hr	144
300,000	2	10 min	0·15
		30 min	0·44
		2 hr	1·8
		10 hr	8·8

analytical columns. For small sample sizes these peak-widths can be assumed to be independent of sample size. It can be seen that, for packed columns ($n \lesssim 10,000$) peak-widths are between 5 and 50 ml, so that for a typical beam area of 0·5 cm^2 a cell-length of between 10 and 100 cm would be required to contain the whole peak and hence maximise sensitivity.

Cells specially designed for GC–IR are available commercially.[10,13] These are of flow-through design, usually with a uniform rectangular cross-section, similar in shape to the incoming beam, which should preferably be focused at the entrance to the cell. However, some of the beam is lost, as it diverges, and to minimise this effect some cells have an internal reflective coating of gold. This also increases the effective path-length of the cell. Cells with tapering sides to enclose the diverging beam have been described by some workers.[14]

Dimensions of three typical commercial cells are shown in Table 10.2, together with estimated sensitivities for detection of weak bands ($\varepsilon = 1\,1\,\text{mole}^{-1}\,\text{cm}^{-1}$), calculated from equation (10.2), using $N/a = 4 \times 10^{-6}$

TABLE 10.2   Characteristics of typical GC–IR gas cells

		Cell A	Cell B	Cell C
Volume	(cc)	20	2·5	0·6
Path-length	(cm)	30	6	6
Approximate sensitivity for weak bands[a]	(μ mole)	3	2	0·4

[a] These figures are based on equation (10.2), with $\epsilon = 1\,1\,\text{mole}^{-1}\,\text{cm}^{-1}$, $(N/a)_d = 4\ \mu\text{moles/cm}^2$, on the assumption that the eluted peak just fills the cell (see p. 212).

mole/cm². The sensitivity for the two largest cells, A and B, is lower by a factor of 5 than equivalent factors for the liquid phase, owing to the large area of the extended beam passing through the sample. This may, how-ever, be offset to some extent by the increased trapping efficiency obtained as compared with condensation techniques.

The smallest cell, C, in Table 10.2, with a volume of 0·6 ml, would only be of use for the most efficient columns (see Table 10.1), unless analysis of only a section of a peak were of interest. The minimum size of sample required to produce a weak band in this cell, ~40 μg, would overload a capillary column, such as the one of 300,000 plates considered in Table 10.1. Thus, even though eluates often emerge from such a column in less than 1 ml of carrier gas, they cannot be readily detected by IR. Support-coated capillaries (sometimes known as porous layer open tube, or PLOT, columns) which are proving increasingly popular may be able to cope with 50-μg samples in peak-widths of about 1 ml, and thus provide sufficient material for the production of a spectrum using cells such as C of Table 10.2.

10.1c On-stream analysis

The continuous monitoring of GC effluents by IR is a recent development. This technique requires that a spectrum be scanned in the short time taken for a peak to emerge from the chromatograph. Inevitably this requirement necessitates some sacrifice of the already rather low sensitivity. A crude but simple continuous technique has been described[15] in which eluates are absorbed in a solution contained in a rotating cell. Successive spectra show more and more features, corresponding to the increasingly complex nature of the mixture in solution.

Good spectra have been obtained (see Figure 10.3) by simultaneously scanning the $2.5—7 \mu$ and $6.5—16 \mu$ regions in 16 seconds.[16] A flow-through by-pass cell of volume 12 ml and length 400 mm was used to trap

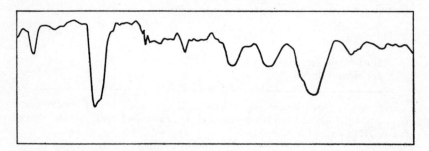

FIGURE 10.3   Infrared spectrum of ethyl alcohol, trapped in the gas phase in a light-pipe after elution from a GC column. Two halves of the spectrum were scanned simultaneously in 16 seconds. Injected sample was $10 \mu l$ of a 10% solution of alcohol in toluene. (From ref. 16)

successive GC peaks. The emergent beam was chopped, alternate signals being applied to two spectrometers, each of which scanned one of the ranges. After a single scan the cell could be purged with nitrogen ready to trap another peak, or further scans of the same sample could be made. Successive peaks separated by more than 32 seconds could be monitored by single scans. A similar arrangement has been described using cell A of Table 10.2, again in a by-pass trapping mode.[17] With a single spectro-meter the range $2.5—15 \mu$ was scanned in 45 seconds. This system has been successfully used for the identification of some constituents of cigarette smoke.[18]

The interrupted elution technique,[14] in which the carrier gas flow is halted while the IR spectrum of an eluted peak is taken, is discussed in some detail on p. 247 in connection with mass spectrometry.

Although all these techniques can, under suitable conditions, produce IR spectra of successive peaks, none involves the use of true on-line monitoring, since the peak to be scanned is isolated in a gas cell. Apart from the limited sensitivity of IR spectroscopy, which was discussed above, the main reason for this is that scans as short as 20 seconds or so are often not really 'fast' in the context of GC. A conventional IR spectrum is invariably obtained by moving a dispersing element (grating or prism), or some associated optical component, so that slit images throughout the wavelength range are focused on the detector. However, movement of a dispersed beam at high speed not only involves mechanical difficulties, but leads to a deterioration in the quality of the spectrum. Thus, it is arguable that conventional IR techniques could never be used for really fast scanning, of 1 second or less, and this probably accounts for the interest in what, to most gas chromatographers, are entirely novel approaches to obtaining IR spectra.

Corner mirrors

One way of avoiding deterioration in spectral quality as a result of fast scanning is to sweep a series of corner mirrors, each consisting of two mirrors at right angles, past a focal plane of the dispersed beam, as shown for a single-beam instrument in Figure 10.4.[19,20] The beam suffers a left-to-right reversal at the corner mirror, and therefore passes back via the stationary grating towards the entrance slit as a doubly dispersed beam. It is interrupted by two mirrors, one for long-wavelength and one for short-wavelength regions of the beam. There is little loss in quality due to the fast scanning itself, and the performance is limited largely by the signal/noise ratio of the detector, as in a conventional slow-scan device. The whole spectrum for the grating used is examined in the time taken for one corner mirror to pass through the focal plane, and this can be as short as 10^{-3} second. In common with other dispersion spectrometers, only a limited wavelength range can be covered by any one grating, e.g. from 2·5 to 9 μ or from 9 to 15 μ.

The use of corner-mirror arrangements has been limited mainly to the study of short-wavelength, transient phenomena, such as those found in stopped-flow kinetics and flash photolysis, but there is no reason why they should not be used for the observation of 'transient' GC peaks.[21] It should be noted, however, that the system remains essentially a dispersion technique, and therefore requires narrow slits, with the accompanying characteristic low sensitivity of conventional IR. However, preliminary experiments have shown that, with a light-pipe similar to cell C of Table

10.2, identifiable spectra of polar molecules such as acetone can be obtained with scan times of 0·5 sec on eluted peaks containing about 25 μg.[22]

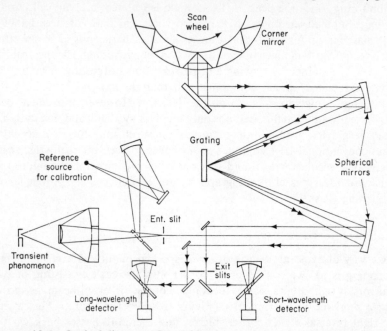

FIGURE 10.4   Optical diagram of a rapid-scanning infrared spectrometer using corner mirrors mounted on a wheel. The short-wavelength path is shown by single arrows, and the long-wavelength path by double arrows. (Courtesy of Warner and Swasey Co.)

Band-pass filters

A non-dispersive spectrometer was available commercially for a short time,[23] which was specifically designed for GC–IR, with a minimum scan time of 5 seconds for the range 2·5—14·5 μ. The heart of this system was a circular variable band-pass interference filter. There was no dispersion, and the optical system was therefore extremely simple, as shown in Figure 10.5. IR radiation from the sample cell passed to the filter, which transmitted only a narrow range of wavelengths to the detector. The wedge-shaped filter rotated, and the corresponding wavelength transmitted changed, so that the detector output imitated that found from a normal dispersion instrument. The scan time was simply the time taken for the filter to rotate. A slit was necessary in order to define that portion of the filter which was being used at any one time. Sample-size requirements were therefore similar to those for dispersion instruments.

Clearly, the principle of operation is very attractive, but doubts have been expressed[5] as to the inherent quality of the spectra produced. Figure 10.6 shows spectra of isopropyl acetate obtained with a filter spectrometer (a) in a 12-second scan on the pure compound and (b) in a 5-second scan on a GC peak containing 0·6 μl. A conventional slow-scan double-beam spectrum (c) is shown for comparison.

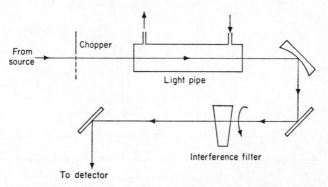

FIGURE 10.5 Optical diagram of a rapid-scanning infrared spectrometer based on rotating wedge-shaped interference filters[23]

Interferometry

IR interferometry is much less commonly used than the dispersion method except for research in the far-IR region beyond 50 μ. However, with suitable interferometers the technique can be used throughout the IR and visible regions of the spectrum.[24,25] A useful introduction to the technique has recently been published.[26]

A typical instrument employs a Michelson interferometer, one mirror of which is moved at constant velocity. When this mirror and the fixed mirror are equidistant from the beam-splitter, both reflected rays of a given wavelength, λ, are in-phase at the detector. This is also true when the moving mirror has been displaced a distance $\lambda/2$, λ, $3\lambda/2$, etc. so that there is a path difference of λ, 2λ, 3λ, etc. for the two rays. However, mirror displacements of $\lambda/4$, $3\lambda/4$, $5\lambda/4$, etc. correspond to path differences of $\lambda/2$, $3\lambda/2$, $5\lambda/2$, etc., so that there is then no signal at the detector for light of wavelength λ. It follows that, for a mirror velocity v, an incoming beam of wavelength λ and intensity I will produce a corresponding interference signal at the detector of frequency $2v/\lambda$, and amplitude proportional to I. If $\lambda = 10$ $\mu = 10^{-5}$ m, i.e. if the incident radiation has a frequency of about 3×10^{14} Hz, the corresponding detector signal for a mirror velocity of

10^{-3} m/sec is a wave of frequency 200 Hz. Thus, the high frequency of the incident IR radiation is transformed to a much lower audiofrequency to which a recorder can respond.

For polychromatic incident radiation, each component wavelength is transformed to a corresponding audiofrequency wave, and the detected

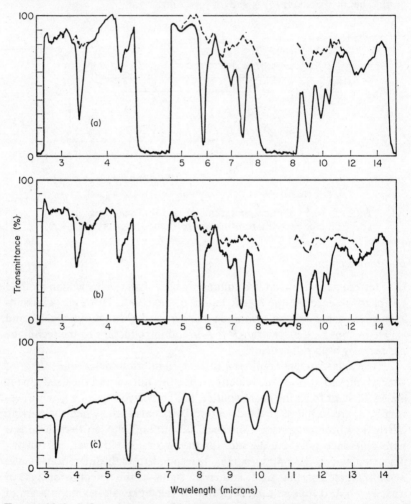

FIGURE 10.6   Infrared spectra of isopropyl acetate. (a) Twelve-second scan of pure compound using the filter spectrometer depicted in Figure 10.5. (b) Five-second scan of GC peak containing 0·6 μl. (c) Slow-scan double-beam spectrum. In (a) and (b) the backgrounds are shown as broken traces, and the spectra are in three sections. (From ref. 23)

signal, known as an interferogram, is the summation of all these low-frequency waves. The interferogram is related to a conventional dispersion spectrum, since it contains information about the intensity of each incident wavelength. In fact, the conventional spectrum is obtained by Fourier transformation of the interferogram. The large amount of tedious calculation involved necessitates the use of additional instrumentation in the form of a wave analyser or a digital computer.

One advantage of an interferometer over a dispersion instrument is that a large aperture of 10 mm diameter or more can be used, so that much less energy is wasted. Furthermore, radiation from the whole wavelength region is incident on the detector all the time. This means that, at a given scan rate and resolution, the signal/noise ratio is very much better than that of a dispersion instrument. At the same time it is easy to cycle the moving mirror in times ranging from 0·1 to 1 second. The relatively high sensitivity and facility for fast scanning are obviously attractive features for GC–IR, and there is commercial interest in this area. The major disadvantage is the high cost and electronic complexity of a scanning interferometer and the equipment necessary to convert the detected interference signal into a conventional % absorption/wavelength spectrum. For an instrument able to scan a spectrum in 1 second with a resolving power of 20 cm^{-1}, the cost in 1969 was about £17,000.

IR interferometry has been used[27-29] to obtain spectra of GC effluents at the 10^{-3}—10^{-4} g level in the region 4—40 μ in 1 second with a resolution of 18 cm^{-1}. An interferometer is available for the more conventional 2·5—16 μ range, which allows X–H stretch bands to be observed. Figure 10.7 shows successive spectra obtained from 1-second scans as an eluted GC peak containing 1 μl of acetone passed through an IR cell 50 mm long and with a volume of 0·6 ml. The growth and decay of the strong bands, e.g. $\nu_{c=0}$, is clearly evident.

The smallest sample size for which this single-scan flow-through mode of operation has been used successfully to date is about 0·2 μl, although the dual-beam technique discussed below allows significant improvement on this. The small cell area of about 0·1 cm^2 does not allow full use to be made of the high light-gathering potential of the instrument, but the corresponding small volume does in principle permit useful information to be obtained about unresolved peaks, since various portions of the peak can be scanned separately. To date, however, this has only been achieved from multiple scans of *trapped* portions of the peaks, using an interrupted elution technique.

Successive scans can be accumulated with the aid of computer storage. Figure 10.8 shows spectra obtained from 100 one-second scans of the front and back of an unresolved peak consisting of 2 μl of iso-octane and methyl ethyl ketone. The change in relative band intensities clearly shows the varying composition of the eluted peak. However, the whole operation

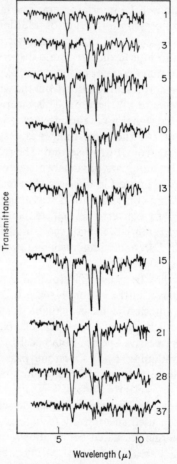

FIGURE 10.7   One-second Fourier transform spectra taken at the times in seconds given beside each spectrum while a GC peak containing 1 μl of acetone passed through the infrared cell. (From ref. 29)

(including subtraction of background spectra) required about 4 minutes, and it seems probable that a conventional dispersion IR could give similar data in about 7 minutes. Thus, if trapping is used, there is little advantage in using interferometry unless the sensitivity of the technique can be made much better than that obtainable with dispersion instruments. A sensitivity limit of about 0·005 μl (i.e. ~5 μg) has been reported for strong bands,[28] which is in fact similar to that found for dispersion methods. However, this limitation in sensitivity is largely due to the use of a 12-bit analogue-to-digital (A/D) converter. Dual-beam techniques[30,31] effectively reduce the dynamic range of the signal, and therefore allow better use to be made of the A/D converter. In conjunction with modern cryogenic solid-state detectors, it is probable that the sensitivity limit will soon be improved to 10^{-7} g.

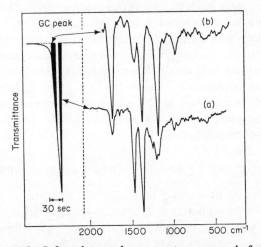

FIGURE 10.8 Infrared gas-phase spectra summed from 100 1-second interferometer scans of two trapped portions of an unresolved GC peak containing 2 μl of a mixture of iso-octane and methyl ethyl ketone. Spectrum (b) is predominantly that of the ketone. (From ref. 28)

Similar results can be obtained by using a single beam and a 14-bit A/D converter. At the same time, this permits spectra to be obtained at higher resolution of 1 cm^{-1} or less, provided that scan times of a few seconds are used and that the position of the moving mirror is known precisely at any instant. This can be achieved by using a subsidiary interferometer which accurately monitors the position of the moving mirror by means of a reference fringe pattern produced by a small laser.[32]

10.2 ULTRAVIOLET SPECTROSCOPY

The general utility of UV is very much less than that of IR, especially in the readily accessible region of about 200—380 mμ, where absorption is largely limited to conjugated systems. This is reflected in the almost total lack of interest in direct coupling of UV and GC. Although the majority of compounds absorb in the far-UV (i.e. down to about 160 mμ), practical difficulties multiply, and in most cases it is questionable whether any more information about structure would be obtained than by IR. Nonetheless, a far-UV spectrophotometer has been coupled to a gas chromatograph, and used to distinguish between all the alkyl iodides containing less than six carbon atoms.[33] However, as has been pointed out,[34] peak identification only becomes a serious problem with *more* than about six carbon atoms, and for these far-UV analysis has low specificity. The spectrometer used in this work was modified to scan 160—210 mμ in 6 seconds, with a resolution better than 0·1 mμ. Although the sensitivity is not quoted, the sensitivity of an earlier design[35] for representative substances was between that of thermal conductivity and flame ionisation. For example, the detection limit for naphthalene, which has $\varepsilon = 1·25 \times 10^5 \, 1\,\text{mole}^{-1}\text{cm}^{-1}$ at 210·8 mμ (very strong, even by UV standards) was about 10^{-8} g. This, of course, is very much better than is possible with IR, because of the much higher molar absorptivity for UV.

10.3 NUCLEAR MAGNETIC RESONANCE

Although NMR is now a vital technique of identification, especially for H, F, B, and P, the weakness of its absorptions means that relatively large samples (~ 1 mg) must be used, even in the so-called micro-tubes, if a reasonably intense spectrum is to be produced from a single scan with, say, a 60 MHz instrument. Computer averaging of repetitive scans can improve the signal/noise ratios considerably. Thus, by scanning over a week-end, usable spectra have been obtained from samples containing 0·3 μmole of hydrogen. This lower limit could probably be reduced by a factor of 5 if a 100 MHz instrument were used.[36] The practical difficulties attending 'on-line' coupling of GC and NMR are immense.

REFERENCES

1. L. J. Bellamy, *The Infra-red Spectra of Complex Molecules*, 2nd edn., Methuen, London (1958).
2. L. J. Bellamy, *Advances in Infra-red Group Frequencies*, Methuen, London (1968).
3. N. B. Colthup, L. H. Daly, and S. E. Wiberley, *Introduction to Infrared and Raman Spectroscopy*, Academic Press, New York (1964).
4. R. M. Silverstein and G. C. Bassler, *Spectrometric Identification of Organic Compounds*, 2nd edn., Wiley, New York (1968), Ch. 3, pp. 64—109.

5. A. B. Littlewood, 'Informal symposium of the gas chromatography discussion group and infrared discussion group', *J. Gas Chromatog.*, **6**, 65 (1968).

6. J. T. Chen, 'Micro KBr technique of infrared spectrophotometry', *J. Assoc. Offic. Agric. Chemists*, **48**, 380 (1965).

7. G. D. Price, E. C. Sunas, and J. F. Williams, 'Microcell for obtaining normal contrast infrared solution spectra at the 5 microgram level', *Analyt. Chem.*, **39**, 138 (1967).

8. Research and Industrial Instruments Co., Worsley Bridge Road, London, S.E.26.

9. N. W. R. Daniels, 'GC-IR microanalysis', *Process Biochem.*, **3**, 34 (1968); and *Column*, **2**, 2 (1967).

10. Wilks Scientific Corporation, 140 Water Street, South Norwalk, Connecticut 06856, U.S.A.

11. A. S. Curry, J. F. Read, C. Brown, and R. W. Jenkins, 'Micro infrared spectroscopy of gas chromatographic fractions', *J. Chromatog.*, **38**, 200 (1969).

12. W. J. de Klein and K. Ulbert, 'A simple micropelleting technique', *Analyt. Chem.*, **41**, 682 (1969).

13. Carle Instruments Inc., 1141 East Ash Avenue, Fullerton, California 92631, U.S.A.

14. R. P. W. Scott, I. A. Fowlis, D. Welti, and T. Wilkins, 'Interrupted-elution gas chromatography. Its application, with eluate concentration, to the automatic production of simultaneous infrared and mass spectra', *Gas Chromatography 1966* (ed. A. B. Littlewood), Institute of Petroleum, London (1967), p. 318.

15. E. A. Haahti and H. M. Fales, 'Continuous infrared functional group detection of gas chromatographic eluates', *Chem. and Ind.*, 507 (1961).

16. A. M. Bartz and H. D. Ruhl, 'Rapid scanning infrared-gas chromatography instrument', *Analyt. Chem.*, **36**, 1892 (1964).

17. P. A. Wilks and R. A. Brown, 'Construction and performance of an infrared chromatographic fraction analyzer', *Analyt. Chem.*, **36**, 1896 (1964).

18. I. Schmeltz, C. D. Stills, W. J. Chamberlain, and R. L. Stedman, 'Analysis of cigarette smoke fraction by combined gas chromatography-infrared spectrophotometry', *Analyt. Chem.*, **37**, 1614 (1965).

19. S. A. Dolin, H. A. Kruegle, and G. J. Penzias, 'A rapid scan spectrometer that sweeps corner mirrors through the spectrum', *Appl. Optics*, **6**, 267 (1967).

20. H. J. Babrov and R. H. Tourin, 'Performance of a commercial rapid scanning spectrometer in emission and absorption', *Appl. Optics*, **7**, 2171 (1968).

21. B. Krakow, 'Continual analysis of gas chromatographic effluents by rapid repetitive infrared scanning', *Analyt. Chem.*, **41**, 815 (1969).

22. S. A. Dolin, H. A. Kruegle, and B. Krakow, 'A rapid-scan spectrometer for continuous flow analysis of gas chromatograph effluents', Pittsburgh Conference on Analytical Chemistry and Applied Spectroscopy, Cleveland, Ohio, March 1968, Paper No. 220.

23. Beckman Instruments Ltd., Glenrothes, Fife, Scotland.

24. G. A. Vanasse and H. Sakai, *Progress in Optics*, Vol. 6 (ed. E. Wolf), Wiley, New York (1967), Ch. 7.

25. J. F. James and R. S. Sternberg, *The Design of Optical Spectrometers*, Chapman and Hall, London (1969), Ch. 8, 'Multiplex spectrometers'.

26. M. J. D. Low, 'Infrared Fourier transform spectroscopy', *Analyt. Chem.*, **41**, 97A (1969).
27. M. J. D. Low, 'Rapid infrared analysis of gas-chromatography peaks', *Chem. Comm.*, 371 (1966).
28. M. J. D. Low and S. K. Freeman, 'Measurement of infrared spectra of GLC fractions using multiple-scan interference spectrometry', *Analyt. Chem.*, **39**, 194 (1967).
29. M. J. D. Low, 'Analysis of gas effluent streams by infrared absorption', in *Gas Effluent Analysis* (ed. W. Lodding), Dekker, New York (1968), p. 155.
30. M. J. D. Low, 'Infrared examination of gas chromatography effluent using a dual beam, single detector interference spectrometer', *Analyt. Letters*, **1**, 819 (1968).
31. P. R. Griffiths, 'Dual-beam Fourier transform spectroscopy', Pittsburgh Conference on Analytical Chemistry and Applied Spectroscopy, Cleveland, Ohio, March 1969.
32. Dunn Associates Inc., 11601 Newport Mill Road, Silver Spring, Maryland 20902, U.S.A.
33. W. Kaye and F. Waska, 'A rapid scan far ultra-violet spectrophotometer for monitoring gas chromatograph effluent', *Analyt. Chem.*, **36**, 2380 (1964).
34. S. G. Perry, 'Peak identification in gas chromatography', *Chromatog. Rev.*, **9**, 1 (1967).
35. W. I. Kaye, 'Far-ultraviolet spectroscopic detection of gas chromatographic effluent', *Analyt. Chem.*, **34**, 287 (1962).
36. R. E. Lundin, R. H. Elsken, R. A. Flath, N. Henderson, T. R. Mon, and R. Teranishi, 'Time-averaged proton magnetic resonance analysis of micro samples from open-tube gas chromatographs', *Analyt. Chem.*, **38**, 291 (1966).

CHAPTER ELEVEN

COMBINED GAS CHROMATOGRAPHY AND MASS SPECTROMETRY (GC–MS)

11.1 LIMITATIONS AND ADVANTAGES OF GC–MS

Useful mass spectra can be obtained only with pure samples, so there is an obvious attraction in using gas chromatography to provide samples for analysis by mass spectrometry. Moreover, the combination of GC and MS is particularly expedient since both techniques can be used with sub-microgram quantities of material. A useful review of the technique covers the important literature up to 1965.[1]

The combination of GC and MS was apparently first described in 1956 when a fraction of the effluent from a packed column was fed into the ionisation chamber of a deflection instrument modified to scan the mass range 12—100 in 60 seconds.[2] The mass spectrum was monitored on a persistent-screen oscilloscope.

During the decade which followed, there were numerous publications by mass spectroscopists describing how to couple a particular instrument to a gas chromatograph.[3–19] At that time, fast scanning was available only with time-of-flight machines, and the scanning and read-out systems of conventional deflection instruments had to be modified to allow scan times of the order of a few seconds. GC–MS has been used almost exclusively by mass spectroscopists, largely because of the relatively high capital and running costs of mass spectrometers compared with the cost of GC itself. As a result, it is now common to find that a laboratory has several gas chromatographs but only, at most, one mass spectrometer. Possible GC–MS configurations are shown in Figure 11.1.

It is only quite recently that a large percentage of gas chromatographers have had the possibility of routine access to a mass spectrometer. Several small, fairly inexpensive, mobile mass spectrometers designed specifically for GC are now available. In addition, several firms now routinely offer highly developed GC–MS units employing deflection, quadrupole, or time-of-flight instruments. Scan times of about 0·1 second are now becoming routine on deflection instruments. Computer systems are available to store and correlate each mass spectrum with the corresponding part of the gas chromatogram. Computerised systems are also being developed to aid identification by comparing experimental and library spectra, and in the case of high-resolution spectra to determine element maps. In short,

GC–MS is becoming big business; the prospect of an analytical machine which can print out the name of each component of a chromatogram is obviously very attractive. It should not be forgotten, however, that quite

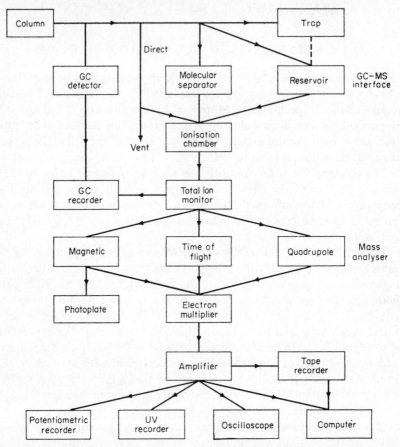

FIGURE 11.1   Summary of the more important GC–MS configurations

apart from economic considerations there are several fundamental limitations to the usefulnesss of MS as an analytical tool, especially when it is used in a fast-scan mode on small samples heavily diluted with carrier gas.

11.1a   Limitations

(*a*) Although the spectra of many known compounds can be understood by hindsight, the interpretation of mass spectra *ab initio* is still in its infancy. Therefore, a mass spectrum must often be regarded simply as a fingerprint, which only permits identification if it can be found in the files.[20-23] The coverage of such files is rather sparse, amounting to only a few thousand compounds, although coverage in catalogues and in the general literature is being extended rapidly. An IR spectrum is similarly useful as a fingerprint, and in some respects permits simpler interpretation in terms of functionality of the molecule. There are about ten times as many documented IR spectra as mass spectra.

(b) Detailed features of a mass spectrum vary considerably from instrument to instrument, even for the same model. It is therefore often essential to confirm an identification by comparing a mass spectrum of the 'unknown' with that of a pure sample of the standard substance. If the mass spectra of unknown and standard are identical, then there is an extremely high probability that the substances are identical. This is not invariably the case, however. For example, mass spectra of cis- and trans-isomers are often indistinguishable. Again, the mass spectra of benzene and butadienylacetylene are identical. Although the mass spectra of positional isomers are frequently quite distinct, differences are often of a minor, quantitative kind, involving only relative peak-heights—and these differences are easily obscured in a fast-scan GC–MS system.

An example is provided by the very similar mass spectra of 3-methylpent-2-ene and 2-ethylbut-1-ene shown in Figure 6.4 alongside the much more characteristic pyrograms. It should be noted that, even if high-resolution fast-scan MS were economically and practically feasible for routine use with GC, such problems would still remain.

(c) The mass spectra of most organic molecules do not include a dominant peak corresponding to the molecular weight of the sample. Indeed, a significant proportion of molecules yield molecular ions which are so unstable that they decompose in the ion chamber.

(d) In GLC, bleeding of the liquid phase from the column can make it impossible to obtain worthwhile spectra, particularly with small samples, unless columns are run at low temperatures (Figure 11.2). Gas–solid columns are not, of course, liable to the same restriction. Even with gas–liquid columns, however, column bleed usually occurs at low mass numbers, and only becomes appreciable at column temperatures in excess of 150 °C, i.e. when high molecular weight compounds are eluted. The high-mass end of spectra may therefore be little affected by column bleed.[16] In some circumstances the MS peaks due to bleed from silicone oil columns can act as useful mass-markers, since they occur at significantly different mass numbers from those of organic fragments.

Bleed effects can be greatly reduced by placing a mini-column of high thermal stability between the column exit and the mass spectrometer. For example, the maximum temperature at which a column could be used was raised in this way from 105 to 190 °C.[24]

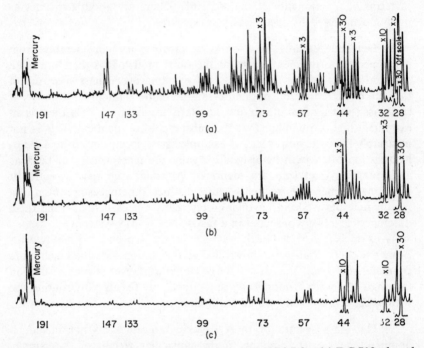

FIGURE 11.2 Mass spectra of column bleed at 200 °C for (a) DC 710 phenyl substituted silicone oil, (b) Carbowax 1540 polyglycol, and (c) Versilube F 50 chlorophenyl substituted silicone oil. (From ref. 16)

(e) Problems of condensation, memory, sample loss, and decomposition associated with the GC–MS interface cannot be lightly ignored. For example, a GC peak containing 4-mercapto-4-methylpentan-2-one $[(CH_3)_2C(SH)CH_2COCH_3]$ was found to give only the mass spectrum of mesityl oxide $[(CH_3)_2C=CHCOCH_3]$ when admitted to a mass spectrometer through a stainless steel jet-type molecular separator (p. 252) at 200 °C. No such decomposition occurred when the separator temperature was reduced to 120 C.[25] Another pitfall which has been observed[26] is the adventitious trimethylsilylation of phenols by traces of bistrimethylsilylacetamide in the spectrometer.

(*f*) The vast dilution of eluate with carrier gas presents real experimental difficulties, which are discussed in detail in section 11.3.

11.1b   Useful characteristics

To set against the limitations listed above we may list the following potential advantages of GC–MS.

(*a*) A molecular ('parent') ion *may* be detected; but since GC cannot easily be used with high-resolution MS, the elemental composition of the eluate may remain in considerable doubt except insofar as peaks containing naturally occurring isotopes may yield important clues. Nevertheless, a knowledge of the molecular weight, accurate to the nearest mass number, is obviously of value in identification of many eluates (see Chapter Seven).

In this connection, it should be mentioned that if a particular molecular ion is too unstable to be detected, it is sometimes possible to produce a more stable ion of molecular weight 1 unit higher by increasing the sample pressure (i.e. injecting larger samples) or by decreasing the repeller potential. With field-emission sources[27] the molecular ion frequently amounts to a very high percentage of the total number of ions produced, sometimes approaching 100%, and usually amounting to at least 10%. However, the use of such sources is practicable only when maximum sensitivity is not required, since sensitivity is typically a hundred times less than with a conventional electron-beam source.

(*b*) In *favourable cases* (and given a skilled interpreter) it is possible to deduce important, and sometimes complete, structural information from detailed consideration of features of the spectrum. Peaks due to rearrangement ions can be useful for characterisation, but can also be very misleading. Metastable ions (i.e. ions which undergo decomposition after leaving the ion chamber but before reaching the detector) give rise to characteristically broad peaks, the positions of which can be of great assistance in structural elucidation. The distribution and intensities of metastables depend to a very large extent on the design of the mass spectrometer.

It should be noted that MS can give information about the position of particular groups in a complex molecule, whereas many other analytical techniques can only indicate whether such groups are present. Several excellent comprehensive texts on the interpretation of mass spectra have been published,[28-32] as well as useful introductory accounts.[33,34]

(c) MS enables spectra to be obtained from sub-microgram amounts of material, thereby yielding fingerprints at levels at which, for example, IR spectrometry, cannot operate.

(d) Fast scanning of the MS at the front and the back of a GC peak makes it possible to check whether or not the peak is due to a single compound. Spectra obtained for different portions of the peak may enable the spectra due to individual components to be distinguished. For example, Figure 11.3 shows the change in intensity of several fragment ions during

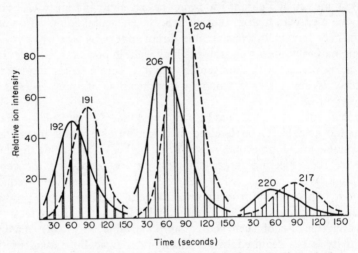

FIGURE 11.3   Changes in intensities of six fragment ions during elution of a mixture of the penta-*O*-trimethylsilyl derivatives of glucose and heptadeuteroglucose from a 3% SE-30 column. (From ref. 35)

the elution of a mixture of trimethylsilylglucose and the corresponding heptadeutero-form from a packed column containing SE-30 silicone oil.[35] Even though the GC peak appeared symmetrical to the eye, it is clear that the deuterated form has a significantly shorter retention time.

(e) Although it may not be possible to identify any particular eluate from its mass spectrum, the appearance of that same spectrum when the sample is analysed on another column makes it a simple, and certain, matter to quote the retention times for an individual component on the two columns, even in a complex mixture. This information may enable the compound type to be determined more surely than by other methods (see Chapter Three), and without the need for re-running of individual peaks or for preliminary class-separation. It is obvious that this use of GC–MS does not require high quality spectra, so that a simple (and relatively cheap) mass spectrometer can be used.

Thus, although MS cannot always provide an unambiguous structure for an unknown compound, it probably provides more useful information than any other single analytical technique, particularly for sub-microgram samples. Nevertheless, in general it will be necessary to supplement the mass spectrum with other information about the eluate, before any confidence can be placed in an identification.

11.2 IMPORTANT FEATURES OF THE MASS SPECTROMETER

11.2a Resolving power

One of the major parameters of a mass spectrometer is its resolving power (R) i.e. the ease with which it can distinguish between ions of similar mass. The most commonly used definition of R is the '10% valley' definition. This involves selecting two peaks of equal height, h, and of masses M and $(M + \Delta M)$ such that the height of the valley between them is $0 \cdot 1h$ (i.e. the depth of the trough between the peaks is $0 \cdot 9h$); then $R = M/\Delta M$. Some workers use a '2% valley' and others a 1 or 2% contribution by one peak to the adjacent peak height as definition of R. None of these definitions readily allows R to be obtained directly from a mass spectrum, since it is unlikely that suitable adjacent peaks will be found. However, for an assumption of triangular peaks the 10% valley definition of R is equivalent to a peak-width of about $0 \cdot 55 \Delta M$ at half-height. Thus,

$$R = M/\Delta M = 0 \cdot 55 M/W_{\frac{1}{2}} \qquad (11.1)$$

Hence, R can be estimated directly from the half-height width, $W_{\frac{1}{2}}$, of a single peak at mass M.

For most deflection instruments, R is approximately constant through-out the mass range, and ideally is determined by the slit-widths. Values quoted for particular instruments usually refer to optimum conditions of slow scan rates and narrow slits, and hence are attainable only with rather large samples. Small samples necessitate faster scan speeds and wider slits, and thus inevitably involve a loss of resolving power. Typical values of R at different scan speeds are shown in Table 11.1 for a variety of commercial instruments. Time-of-flight instruments and the increasingly popular quadrupole instruments are included in Table 11.1, although the factors affecting R are different from those operating in deflection spectro-meters. In particular, the width at half-height of mass peaks obtained from quadrupole instruments has a constant value of about half a mass unit over the whole range, so that the resolving power as defined in equation (11.1) is approximately equal to the mass of the peak being scanned.

With time-of-flight and quadrupole instruments, unit mass separation up to about 500 can be obtained. This is quite adequate for most GC–MS work, since it allows determination of the mass numbers of the heaviest ions in a spectrum to the nearest unit. For many systems a value of R of

TABLE 11.1   Performance characteristics of typical mass spectrometers

Instrument	Resolving power	Fastest corresponding scan time per decade	Approximate cost, including inlet, read-out, and pumps (1969)
Magnetic deflection (permanent magnets)	200 200	10 min 10^{-1} sec[a]	£5000 £7500
Magnetic deflection (electromagnets)	5000 or 1500 or 500	10 min 5 sec 1 sec	£20,000
Double focusing	800	5 sec	£15,000
Double focusing	50,000 or 10,000 or 1000	10 min 10 sec 2 sec	£35,000
Time of flight	300	10^{-4} sec[a]	£15,000
Quadrupole	[b]	10^{-1} sec[a]	£10,000

[a] These are typical repetition rates which can be displayed on an oscilloscope. The fastest corresponding recorder output would take about 1 second.

[b] Quadrupole instruments (mass filters) scan linearly and produce mass peaks of equal width (about 0·5 mass unit at half-height) up to about mass 500. Thus, the conventional resolving power is approximately equal to the mass number being scanned [see equation (11.1)].

about 200, as provided by the small, deflection instruments, would suffice. For determination of the elemental composition of a molecule, however, it is necessary to know the mass number of each ion to at least 0·005 of a unit, which requires a resolving power of 10,000 or more. This can be achieved only with the expensive double-focusing (magnetic plus electric) deflection instruments, with scan times no smaller than about 10 seconds.

11.2b Scan rate

Scanning is achieved in a variety of ways in different instruments. That most commonly used in *deflection spectrometers* is the variation of the magnetic field strength at constant accelerating voltage. Much faster scanning with deflection instruments can be achieved by varying the accelerating voltage at constant magnetic field, and this is unavoidable with the small permanent-magnet instruments. Some of the larger spectrometers have facilities for using magnetic scanning for fairly slow scans of a few seconds or more, and voltage scanning for fast scans of 1 second or less. However, voltage scanning can be used only over a limited mass range, and also leads to distortion of relative ion intensities compared with magnetic scanning.

Voltage scanning of a different kind is usually used with *quadrupole instruments*, again allowing scan times of less than 1 second. The amplitude, V, of a fixed-frequency a.c. voltage is varied, and the mass/charge ratio of ions which undergo stable trajectories through the filter without being removed at the quadrupole rods is then proportional to V. As mentioned above, the effect of this scanning procedure is that the separation of adjacent mass peaks is constant, in contrast to deflection-instrument spectra where the absolute separation of adjacent peaks decreases steadily with increasing mass number.

Time-of-flight instruments have an inherently very fast 'scan rate', since flight times are of the order of a few microseconds. Thus, as many as 10^5 complete spectra can be obtained per second. This is, of course, too fast for all direct read-out systems except an oscilloscope. For an analogue output such as an oscillographic recorder, it is therefore necessary to slow the real-time presentation of the spectrum by sampling progressively later stages of successive flight pulses.

If the mass range is scanned at a rate of $\mathrm{d}M/\mathrm{d}t$ mass units per second, then for small increments we have

$$\Delta t = \Delta M/(\mathrm{d}M/\mathrm{d}t)$$

where Δt is the time to scan a peak of width ΔM at mass M. Since $R = M/\Delta M$ [equation (11.1)], it follows that

$$\Delta t = \frac{M}{R} \bigg/ \left(\frac{\mathrm{d}M}{\mathrm{d}t}\right) \qquad (11.2)$$

Quadrupole instruments employ linear scanning ($\mathrm{d}M/\mathrm{d}t$ constant) and R is approximately equal to M, so that Δt is constant, and therefore all peaks are of equal width.

For deflection instruments, however, for which the resolving power R is constant, it follows from equation (11.2) that, if linear scanning is used, Δt will increase linearly with M so that peaks will become progressively broader. Thus, peaks near $M = 2$ would require a recorder response one-hundred times faster than peaks near $M = 200$. It is much more usual, therefore, to employ *exponential scanning* for deflection instruments, so that the scan rate, dM/dt, increases linearly with M. Hence, $dM/dt = kM$, and

$$M = M_0 \exp(kt) \tag{11.3}$$

where M_0 is the mass at which scanning begins at zero time, and k is constant. It follows from equation (11.2) that

$$\Delta t = 1/Rk \tag{11.4}$$

Hence, with exponential scanning at constant resolving power, equal time is spent scanning each mass number. This means that best use can be made of the capabilities of the amplifier and recording systems. For a constant chart speed, peaks from deflection instruments will then be of equal width at all stages of the exponential scan, although the separation between the tops of adjacent mass peaks will decrease at higher values of M. This makes determination of mass number more difficult than is the case with linear scanning which involves equally spaced peaks.

For an exponential scan we obtain, from equations (11.3) and (11.4),

$$t = R\Delta t \ln(M/M_0) = 2 \cdot 303 R \Delta t \log_{10}(M/M_0)$$

It is usual to express exponential scan rates in terms of the time, t_{10}, required to scan a mass decade, i.e. for $M/M_0 = 10$. From the above equation we then have

$$\Delta t = \frac{t_{10}}{2 \cdot 303 R} \tag{11.5}$$

This useful expression, which also applies when scanning from high to low mass, relates the time needed to scan a single peak, Δt, to the scan rate and the resolving power of the instrument. In GC–MS it is essential to make t_{10} smaller than the time of elution of the GC peak. A useful rule-of-thumb is that the time taken to scan the portion of the spectrum of interest should not exceed one-fifth of the width of the GC peak at half-height.

Even so, it is difficult to ensure that the scan is taken round the top of a peak, where the eluate concentration is changing least rapidly. Changes in concentration of eluate in the ion chamber during scanning lead to distortion of the mass spectrum, so that no reliance can be placed on relative peak-heights in identification. Such distortion can be removed by ratio recording, i.e. by recording each individual ion current as a fraction of the instantaneous total ion current. Contribution due to the carrier gas can either be backed off or eliminated by the use of a low accelerating voltage in a secondary ion chamber or in the chamber proper. Ratio recording can be done with a computer or with a measuring bridge.[36] A less expensive solution is illustrated in Figure 11.4.[37] The output from the electron

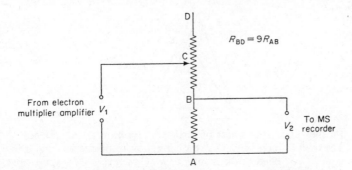

FIGURE 11.4   Schematic illustration of a device for ratio-recording mass spectra. Potentiometer wiper C is driven by the total-ion monitor recorder. See text. (From ref. 37)

multiplier amplifier, V_1, is placed across a fixed resistor AB in series with a potentiometer BD, the wiper of which, C, is mechanically coupled to the slide-wire of the total ion current recorder. C is positioned at B and D for total ion currents of 10% and 100% full scale, respectively, and R_{BD} is exactly $9R_{AB}$, so that R_{AC} is directly proportional to the total ion current, i.e. $R_{AC} = kI_{tot}$. The voltage across AB, V_2, is fed to the mass spectrum recorder and is given by

$$V_2 = \frac{R_{AB}V_1}{R_{AB} + R_{BC}} = \frac{R_{AB}V_1}{kI_{tot}}$$

Since V_1 is a measure of the individual ion current being scanned, it follows that V_2 is a measure of the ratio of individual ion current to total ion current for all values of I_{tot} between 10% and 100%. The recorded spectrum will be equivalent to one obtained at a constant total ion current of 10% full scale.

11.2c   Amplifier and recorder requirements

There is an inevitable loss in resolving power (increased peak-width) and sensitivity (decreased peak-height) as a result of fast scanning, as shown schematically in Figure 11.5 for a triangular peak. A detailed treatment of this problem[38] shows that the relative decrease in peak-height is given by

$$\frac{h-h'}{h} = \frac{2\tau}{\Delta t} \ln\left(2 - e^{-2\Delta t/\tau}\right)$$

where τ is the time constant of the recorder–amplifier system. It follows that, if there is to be less than 5% loss in sensitivity as a result of fast

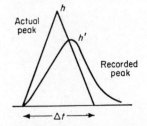

FIGURE 11.5   The effect of fast scanning on a triangular peak. The width increases, i.e. resolving power decreases, and the height falls from h to h'. (From ref. 38)

scanning, it is necessary to have a recorder–amplifier system with τ less than $0{\cdot}036\Delta t$. Thus, from equation (11.5) we have

$$\tau < 0{\cdot}036\Delta t \approx \frac{t_{10}}{64R}$$

while the required band-width, B, of the system in Hz is given by

$$B = \frac{1}{2\pi\tau} \gtrsim \frac{10R}{t_{10}} \tag{11.6}$$

As expected, this equation shows that high resolving power and fast scan rates require output systems with large band-widths. The minimum band-widths required for various scan speeds and resolving powers as calculated from equation (11.6) are given in Table 11.2, while Figure 11.6 shows the

TABLE 11.2   Required bandwidths of amplifiers and recorders (Hz) calculated from equation (11.6)

Resolving power	Scan time per decade		
	1 sec	1 min	1 hr
50,000	500,000	8000	140
10,000	100,000	1700	28
800	8000	130	2
300	3000	50	0·8

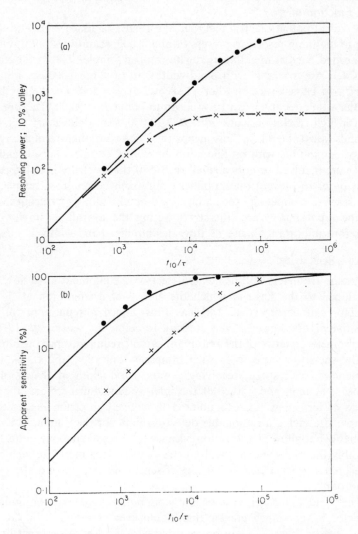

FIGURE 11.6 The theoretical and experimental effect of fast scanning on (a) resolving power and (b) peak-height. t_{10} is the exponential scan time for a decade in mass, and τ is the time constant of the amplifier–recorder system. ●, Static resolving power 8000; ×, static resolving power 590. (From ref. 38)

theoretical and experimental variation of resolving power and sensitivity as a function of t_{10}/τ.[38]

A typical potentiometric recorder has a band-width of 5 Hz and therefore to maintain resolving powers of about 300, scan times of 10 minutes are needed, such as might be used with trapped samples. UV oscillographic recorders are available with band-widths up to about 5000 Hz, allowing spectra to be recorded in about 1 second at $R \approx 500$. For higher band-widths a tape recorder can be used up to about $B = 50,000$ Hz, but even so, for high-resolution work with $R = 10,000$, scan times must be several seconds (see Table 11.2). This means that 'fast scan' high-resolution work is not compatible with on-line coupling with GC unless very broad GC peaks are involved or only a small region of the spectrum is of interest.

In order to prevent the amplifier noise from becoming excessive, it is necessary to compensate for an increased band-width by reducing the size of the input impedance, thereby lowering the overall sensitivity. It is therefore important to use as low an amplifier band-width as possible.

11.2d   Sensitivity

Mass spectrometers and associated electronic equipment have now been developed to the stage at which the arrival of a single ion at the MS detector can be recorded. MS has thus gained a reputation for high sensitivity. It is, however, very difficult to define the sensitivity of a mass spectrometer in terms of the amount of sample required, since the intensity of the spectrum varies with such things as ionisation and transmission efficiencies, scan speed, resolving power, band-width of the amplifier–recorder system, and nature of the inlet system. Furthermore, the wide range of ion intensities encountered in a given spectrum means that a sample size giving a reasonable detection limit for the dominant ion will probably not allow identification because the overall cracking pattern and possibly the molecular ion will not be evident. It is therefore possible to give only a general idea of the sort of sensitivity attainable under typical conditions.

The potential sensitivity of the commonly used electron-multiplier detector is very much greater than could ever be required by GC–MS. Thus, reasonable peaks can be recorded with a lower ion current limit of about 6×10^{-20} amp, which corresponds to the arrival of one ion every second, so that statistical effects are important. In order to record such a spectrum, the time to scan each mass peak, Δt, must be about 100 seconds. At a modest resolving power of 300 it follows from equation (11.5) that it would then take 20 hours to scan a mass decade. With the same overall sample consumption, an essentially similar spectrum could be recorded in a few seconds by increasing the rate of inflow of sample and thereby increasing ion currents. Thus, detector sensitivity is never a limiting factor.

If we set a requirement that ten ions must be collected while scanning the weakest peak in time Δt, then the ion current at that peak must correspond to $10/\Delta t$ ions per second. If this weakest peak is 0.1% of the total ion intensity, it follows that $10^4/\Delta t$ ions per second must arrive at the detector in total. If we define an overall efficiency, E, for the conversion of molecules in the ion chamber into ions at the detector, then $1/E$ molecules per second must pass through the ion chamber to produce 1 ion per second at the detector, so that the required sample flow rate is $10^4/E\Delta t$ molecules per second. The total amount of sample consumed during the scanning of a mass decade, assuming a molecular weight of 100, is therefore $10^{-17} t_{10}/6E\Delta t$ gram. Using equation (11.5) this can be rewritten approximately as $10^{-17} R/3E$ gram per decade which depends only on the resolution and efficiency, and not on the scan rate—except insofar as very fast scanning leads to a reduction in R. A typical value of E for moderate resolution is about 10^{-5}, so that at $R = 300$ the total sample requirement is about 10^{-10} g per decade, whatever the scan speed.

For high-resolution spectra the number of observed peaks is much greater than for low-resolution work, so that it is more reasonable to set the weakest peak as 0.01% of the total intensity. The expression for the total sample requirement then becomes $10^{-16} R/3E$ gram per decade. The narrow slits required for $R = 30,000$ correspond to a greatly reduced value of E, of say 10^{-8}, so that about 10^{-4} g per decade would be required. In order to achieve such high resolution, however, slow scanning would be essential. Table 11.3 shows theoretical sample sizes and scan-time requirements which enable mass measurement of better than 1 in 10^5 at high resolution, on the assumption that 10^{-7} g/sec of sample enters the ion source.[39]

TABLE 11.3 Sample requirements for high-resolution mass spectrometry (see text; cf. ref. 39)

Resolution	Scan time per decade (sec)	Sample consumption (g)	Band-width (Hz)
10,000	10	10^{-6}	10,000
20,000	40	4×10^{-6}	5000
30,000	1000	10^{-4}	300

With electrical detection, the sample requirements discussed above can be decreased only by sacrificing the quality of the spectrum or by using a more efficient mass spectrometer. Photographic recording can be used in the Mattauch–Herzog type of deflection spectrometers, in which all ions are focused in one plane. It is then not the ion current which is detected, but the total number of ions arriving during the exposure. The sensitivity limit for a photo-plate is about 10^4 ions, so that about 10^7 ions would be required for a reasonable spectrum, if again we assume that the weakest peak represents 0·1% of the total intensity. With $E = 10^{-5}$, as assumed for electrical detection at $R = 300$, this sets the lower limit at 10^{12} molecules of sample per decade, or about 10^{-10} g, which is similar to the sensitivity for electrical detection at moderate resolution. For a very high resolution spectrum requiring 10^8 ions with $E = 10^{-8}$, however, 10^{-6} g would be required, which is two orders of magnitude more sensitive than with electrical detection under similar conditions. In practice the smallest sample sizes reported for reasonable photo-plate spectra at high resolution are about 10^{-8} g,[40] compared with between 10^{-6} and 10^{-7} g for a 10-second scan at $R = 10,000$ with electrical detection.[39]

In summary, it is reasonable to expect to be able to obtain fast-scan low-resolution spectra from nanogram samples, and high-resolution spectra from microgram samples.

11.3 SAMPLE TRANSFER FROM GC TO MS

There are two general methods of introducing samples from a gas chromatograph to a mass spectrometer. The first involves collecting the eluate in a suitable reservoir, and then obtaining the mass spectrum at leisure. The second involves direct coupling of the column exit to the ion chamber so that the mass spectrum must be obtained while the GC peak is actually emerging. Both techniques have attractive features as well as handicaps, so it is not possible to say that one approach is better than another except in the context of a particular problem.[41]

11.3a Trapping techniques

The simplest way of obtaining the mass spectrum of an eluate is to collect that portion of carrier gas which contains the peak of interest, and to transfer this to the mass spectrometer reservoir. However, because of the large excess of carrier gas, it is necessary to operate at a much higher source pressure than would be required for the same amount of eluate in the absence of carrier gas. Since the pressure in the ion chamber must not exceed about 10^{-4} mm Hg, this will often limit the rate of sample input to the source, with a consequent reduction in the intensity of the spectrum. It is therefore often advantageous to freeze the eluate and pump off the

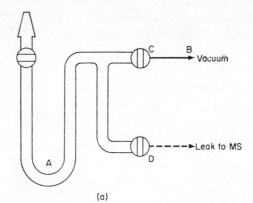

(a)

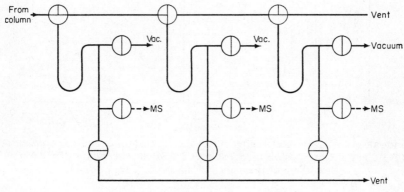

(b)

FIGURE 11.7   (a) A small-volume trap–reservoir suitable for introducing GC peaks to a mass spectrometer. (b) A multi-trap system for sequential collection of GC peaks. (Cf. refs. 9 and 42)

carrier gas before running a spectrum. This is usually not possible for samples of less than about 10^{-4} g, for which efficient cold-trapping is very difficult to achieve (cf. Chapter Nine).

Most mass spectrometers are provided with reservoirs of 1 litre capacity for sample introduction, in order that the observed mass peaks will not be affected to any significant extent throughout a spectrum by sample depletion. However, expansion of the contents of a GC peak into a volume of this size results in a significant fall in sensitivity. For example, a sample of 10^{-6} g would exert a pressure of only about 10^{-3} mm Hg. Therefore, for all but milligram samples, it is desirable to provide a reservoir with a volume of a few ml, as indicated in Figure 11.7(a). Typically the tube A is a trap, cooled to reduce the vapour pressure of the sample to less than

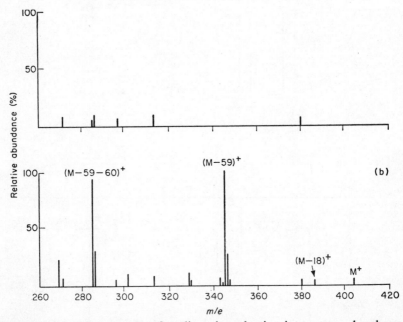

FIGURE 11.8 Mass spectra after direct introduction into source chamber of (a) 1·5 mg of Chromosorb P coated with 5% SE-30 silicone oil; (b) the same, plus 10^{-7} g of pregnanediol diacetate. (From ref. 40)

10^{-5} mm Hg. After evacuation of air and carrier gas through B, tap C is closed and trap A is warmed to evaporate the eluate. The mass spectrum of the eluate, relatively free from carrier gas, is then obtained by adjustment of the restrictor D to allow a suitable leak rate into the source chamber. Ratio recording (p. 239) can be used to allow for the effect of sample depletion on the observed peak heights.[37]

Figure 11.7(b) shows an inlet system which allows sequential collection of GC peaks, one in each trap, for subsequent analysis.[9,42] Clearly, the use of such systems can become cumbersome, and introduces dead-space and temperature-control problems. Since there is bound to be a delay between completion of a GC run and running of all the required mass spectra, it may often be possible to achieve just as rapid operation by using multiple chromatographic injections, and trapping a different eluate in each run.

A very attractive technique for introducing trapped samples directly into the source chamber has been described,[40] in which samples are trapped in mini-columns of the type described on p. 203. These consist of capillary tubes filled with coated column packing, which are attached to the column exit and held below column temperature, thus retaining a peak with a high efficiency. After such trapping, the capillary can be inserted in a solid sample probe before being introduced into the ion chamber. These devices have been used successfully for obtaining mass spectra of GC peaks containing as little as 10^{-7} g of eluate. Figure 11.8 shows the mass spectrum of 10^{-7} g of eluate obtained in this way.[40] Activated charcoal has similarly been used for trapping small amounts of volatile eluates prior to mass spectrometric analysis.[43]

11.3b   Interrupted elution

The trapping techniques discussed above are applicable to any normal gas chromatograph. In such a system, any peak which is not trapped is lost. By halting the carrier gas flow, however, it is possible to store peaks in the GC column itself while the peak which has just emerged is collected in a reservoir, so avoiding the need for repeated injections. After a mass spectrum has been taken, the carrier gas flow can be restarted until the next peak has been eluted. Such a stop–go system was used in early work[6] where the carrier gas was removed in a freeze–pump–thaw cycle, as discussed above.

More recently a rather sophisticated automatic stop–go system has been used for IR as well as mass spectrometry.[44] A simplified block diagram of the apparatus is shown in Figure 11.9. Under normal conditions all taps except A, D, and E are closed, as indicated in the diagram. Thus, carrier gas flows through the flow controller and the pre-column before passing the injection port and entering the main column. Most of the effluent passes via taps D and E to vent, but a fraction (say 10%) flows through the flame detector. Note that the trap and associated capillary tubing form a cul-de-sac full of stationary carrier gas. The automatic sequence of operations is started by a deflection of the recorder as a peak begins to emerge, and is as follows.

(a) *Trap peak*. Tap E closes and F opens so that effluent is diverted through the trap. This consists of a short column held at low temperature. It must be long enough to hold eluates with the shortest retention times.

I

Usually 1—2 inches of column should suffice. When the recorder pen reaches the tail of the peak, tap F closes and E re-opens. Simultaneously, however, the flow through the analytical column is halted, and the trap is heated.

(b) *Stop flow*. Taps A and D close, the pressure in the column is *slowly* released at C, while the flow controller is vented at B. In this way, peaks undergo backward development, and since the pressure near the column

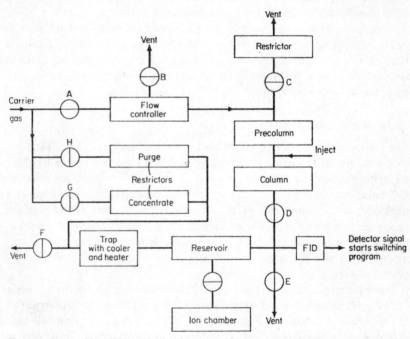

FIGURE 11.9   Simplified block diagram of an automatic interrupted elution system. (Cf. ref. 44)

inlet is much lower than normal, peaks here may move further back than they have come. It is for this reason that the pre-column is essential. Note that if the flow were stopped simply by closing D, then forward chromatographic development would continue until the whole column was at the inlet pressure. This could lead to an undesirable sudden pressure change at the column exit on re-starting the flow, and resolution could be impaired. However, it has been reported that this simple procedure can give just as satisfactory results as the pressure-release method.[45]

(c) *Concentrate eluate.* When the trap has been heated for about 2 minutes, tap G opens and the contents of the trapping column are back-flushed into the reservoir. This can be an IR cell. Because of the increased temperature, the volume of carrier gas required to remove the eluate from the trap will be considerably smaller than that used to place it on the cold trap. In this way concentration by a factor of 10 or more can readily be achieved.

An obvious defect of the system described is the presence of the reservoir volume (an IR cell) between the trap and the analytical column. A more satisfactory arrangement would have a capillary by-pass to allow direct transfer from D to the trap.

(d) *Take mass spectrum.* When the peak has been back-flushed into the reservoir, tap G closes and the trap heater is turned off. The mass spectrum is then scanned over a period of about 10 minutes.

(e) *Purge trap.* The trap heater is turned on again, and the trap and reservoir are purged by a fast flow of gas through tap H for 1 or 2 minutes, venting taking place through E. Tap H then closes, the heater is turned off, and cooling begins.

(f) *Start flow.* Taps B and C close, A and D open, but the flow controller allows only a slow build-up of pressure over about 5 minutes at the column inlet. The recorder signal accompanying the front of the next peak trips the whole sequence again.

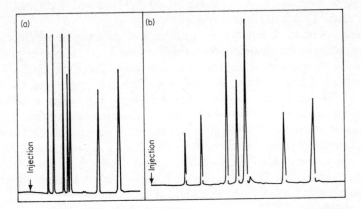

FIGURE 11.10 Chromatograms of a mixture of fatty acid ethyl esters obtained (a) without and (b) with interrupted elution. (From ref. 44)

I*

Using a 15-foot pre-column and a 60-foot analytical column with flow rates of 100 cc/min, this system has been used to obtain mass spectra of upwards of 10 μg of eluate. If the whole peak were to be contained in only a few ml of carrier gas (cf. Table 10.1) there would be no need to carry out the peak-concentration procedure, and a simple total-effluent trap would suffice (see p. 205).

As can be seen from Figure 11.10, there is only slight peak-broadening as a result of the halted flow. The apparently increased distance between peaks is not real; it results from the fact that the recorder is switched off when the flow is stopped, but is switched on again at the beginning of the slow restoration of inlet pressure.

11.3c Sample-enrichment devices

If x ml of carrier gas (corrected to $0\,°C$ and a pressure of 1 atmosphere) are eluted between the emergence of the front and back of a triangular GC peak containing N moles of material, then the average percentage of eluate in the carrier gas is about $2 \times 10^6\,N/x$. For the typical GC peaks emerging from a packed column, this is in the range $10^4\,N\%$ to $10^6\,N\%$ (Table 10.1). For a molecular weight of 100, a 'large' GC sample size of 10^{-3} g corresponds to 10^{-5} mole, and hence to an abundance of between 0.1% and 10% of the carrier gas. For a 10^{-6} g sample, the corresponding abundance would only be about 0.01—0.0001%.

For a capillary column, the 10^{-6} g sample would be more concentrated, amounting to between 0.1% and 0.01% of the effluent, but the column would be overloaded. A more typical 10^{-9} g load would correspond to only about $10^{-4}\%$ of the carrier gas flow.

Such enormous dilution is no drawback to direct mass spectrometric examination of the eluate provided that the large excess of carrier gas does not raise the pressure in the ion source chamber above the limit for safe operation, which is about 10^{-4} mm Hg. Thus, the maximum tolerable rate of flow of carrier gas plus sample into the ion source chamber is governed by the speed at which the ion chamber is pumped. This is usually in the range 30—300 litres/sec, so that a pressure of 10^{-4} mm Hg in the source chamber corresponds to an inflow of between about 0.2 and 2 cc/min at atmospheric pressure. Thus, most of the effluent from a capillary column can be allowed to enter the source chamber, but only 5% or less of the effluent from a packed column. It follows that direct coupling of packed columns with a mass spectrometer must involve the loss of at least 95% of the maximum possible sensitivity for the sample size used, however small. It is for this reason that interest has developed in the use of molecular separators, which, by concentrating the eluate, allow a greater total percentage of the sample to enter the source.

With most separators the enrichment factor is only about 100, so that carrier gas molecules entering the source chamber still commonly out-number eluate molecules by 1000 or more to 1. However, this modest enrichment can allow almost all the eluate from packed columns to enter the mass spectrometer.

Fritted-glass tube

A commonly used molecular separator is of the fritted-glass type first devised by Watson and Biemann,[46] as shown in Figure 11.11. This device is normally used with helium carrier gas, which preferentially diffuses

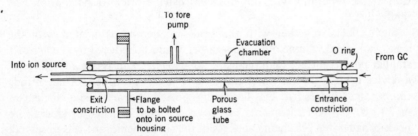
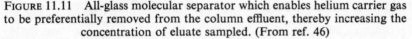

FIGURE 11.11 All-glass molecular separator which enables helium carrier gas to be preferentially removed from the column effluent, thereby increasing the concentration of eluate sampled. (From ref. 46)

through the sintered glass wall of the inner tube and is pumped away. The 'ultrafine porosity' fritted tube is about 20 cm long with an outside dia-meter of about 7 mm, while the outer vacuum chamber has a diameter of about 3 cm. With the outlet of the fritted tube sealed off, the inlet restrictor should be chosen to give a pressure in the vacuum chamber of about 1 mm Hg under normal conditions of flow rate and pumping. An exit restrictor capillary, 5 cm long with a 0·3 mm bore, should then give an acceptable leak rate to the source chamber.[47] Sample enrichments of 50—100-fold have been reported.

With this sort of separator a steadily increasing fraction of the eluate will diffuse out of the tube as its concentration increases. The total fraction of eluate entering the mass spectrometer is therefore usually limited to between 10 and 30% for flow rates round 30 ml/min. Under these condi-tions carrier gas enters the source at the rate of about 0·2 ml min^{-1}.

There is a danger with any interface system that surface effects will cause loss of efficiency, tailing, or even complete loss of sample. The fritted-glass type of separator can give good transmission from column to mass spectrometer for microgram quantities provided that care is taken to keep the whole unit at column temperature. However, several workers have noted that very small samples (10^{-9} g) can suffer tailing and significant sample loss. For example, no mass spectrum trace could be obtained for samples of terpenoid alcohols and aldehydes of less than 10^{-6} g, although ethers and olefins could be detected at the 10^{-9} g level.[48] Similar behaviour towards high-boiling esters has been reported.[49] In view of the common use of directly coupled GC–MS for nanogram samples, this situation is potentially very serious. However, it was found that after *in situ* silanisation of the separator with bistrimethylsilylacetamide, 10^{-9} g of all types of terpenoids could be detected. It should be noted, however, that silanisation of the separator with dimethylchlorosilane prior to installation was ineffective.[48]

Such effects, together with the peak-broadening associated with the significant dead volume of the separator, would be reduced by the use of a smaller porous surface than the usual 25—50 cm². This is one advantage of a device using a 0·1 cm² *silver membrane*, some two orders of magnitude thinner than a glass frit. This has been reported to give transfer efficiencies of 30—60% with enrichments of between 5 and 25 while producing no peak-broadening. No chemical effect of the silver membrane was observed, even with halides and sulphides at 240 °C.[49]

Perfluoro-polymer membrane

Thin-walled capillary tubes made of a copolymer of C_2F_4 and C_3F_6 (Teflon FEP) have also been used to allow selective removal of helium carrier gas by permeation.[50] A temperature of about 250 °C is required, and quoted transfer efficiencies are around 50% for a 7-foot tube with walls of 0·1 mm thickness. Although this system allows a rather simpler inlet arrangement, workers who have compared it with the fritted-glass tube preferred the latter.[51] It should be noted that the use of a polymer of C_2F_4 (PTFE) as a GC–MS interface is undesirable if silanising solutes such as trimethylchlorosilane and bistrimethylsilylacetamide are used, since the PTFE is attacked to give trimethylfluorosilane.[52] It seems probable that the FEP copolymer would react similarly.

Jet-diffusion

Another device which makes use of the preferential diffusion of helium carrier gas molecules to achieve sample enrichment is shown schematically in Figure 11.12. Effluent at atmospheric pressure enters the stainless-steel separator at A and passes through the jet B at supersonic velocity. There is a preferential removal of the light helium carrier gas by diffusion in the evacuated gap C before the enriched sample enters the orifice at D, which is exactly aligned with B. After passing through a similar second stage the sample enters the mass spectrometer at F.

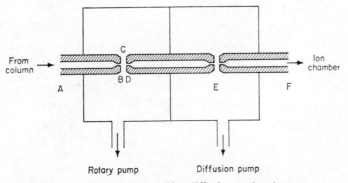

FIGURE 11.12　Two-stage metal jet-diffusion molecular separator, operating in a similar manner to the device shown in Figure 11.11. (Cf. ref. 15)

Based on a design by Becker,[53] this dual-jet system has been developed by Ryhage and co-workers to connect packed columns directly to an LKB commercial mass spectrometer.[15] A pressure of 0·1 mm Hg is produced after the first stage of the separator at C by a rotary pump, while the pressure of the second stage at E is reduced to about 10^{-4} mm Hg by a fast diffusion pump which removes the need for differential pumping of the source chamber. Typical enrichment factors for this separator are about 80 (very similar to that for the fritted-glass type), and it appears that about 20% of the eluate can be transferred to the source if the separator is maintained at column temperature.

Membrane solubility

All the molecular separators discussed above work on the principle of preferentially removing carrier gas molecules from the effluent. The converse procedure, i.e. the selective removal of solute molecules from the effluent, is the basis of the two-stage separator developed by Llewellyn.[54] This device, which is available commercially, is shown in Figure 11.13. The permeable methyl silicone rubber membranes (about 0·02 mm thick) have a very much greater affinity for organic solute molecules than for carrier gas molecules, and very high enrichment factors of the order of 10^5 are possible at flow rates round 50 cc/min. As indicated earlier, however, the important thing is the overall transfer efficiency, and this is reported to be around 50%, i.e. of the same order as that of other separators.

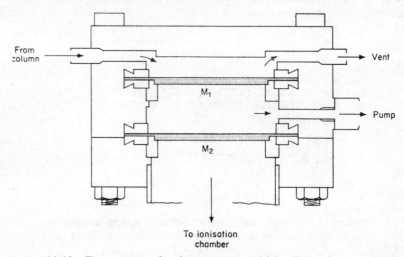

FIGURE 11.13   Two-stage molecular separator which allows eluate molecules to diffuse preferentially through the rubber membranes, M_1 and M_2. (Courtesy of Varian Associates)

The high enrichment factor does, however, mean that only a simple pumping system is required. For microgram samples the ratio of carrier gas to eluate molecules entering the source will be about 1, and there is no need for differential pumping of the ion source, since the pressure will be below 10^{-6} mm Hg even at pumping speeds as low as 1 litre/sec. Furthermore, in contrast to the other separators, the column exit is run at atmospheric pressure.

With this separator there is no need to use a 'light' carrier gas, and in fact the membrane has a rather lower conductance for nitrogen than for helium. A possible drawback to very fast scanning with this separator is the time required for passage through the membrane (1 or 2 seconds). Serious band-spreading has been reported, although this can be improved to some extent by suitable temperature-programming of the separator.[55] The transfer time can be decreased by use of only one stage of the separator, but the methyl silicone polymer is too permeable to carrier gas to allow all the material transferred through a 3 cm^2 membrane to enter the mass spectrometer. However, a single-stage version has been constructed using a membrane of 60% phenyl, 40% methyl siloxane copolymer painted on a porous silver membrane. This is much less permeable to carrier gas than the methyl silicone, and all the material passing through a 6 cm² membrane could be handled by a mass spectrometer with a pumping speed of 160 l/sec, giving a source pressure below 5×10^{-6} mm Hg. Peak-broadening was still apparent, but was less severe than with the two-stage device. Overall transfer efficiencies of about 40% were obtained.[55]

The fact that the separator is selective to compound type opens up the interesting possibility of using GC–MS as a selective dual detector by comparing the total mass spectrometer ion current with the output of a GC detector.

11.3d   Direct coupling with capillary columns

As indicated on p. 250, up to about 2 ml/min of gas at atmospheric pressure can enter a mass spectrometer source chamber being pumped at 300 l/sec before pressures in excess of the safety limit of 10^{-4} mm Hg are reached. Since atmospheric flow rates in 0·01-inch capillary columns are about 1 ml/min, it follows that such columns can be coupled directly to a mass spectrometer without the need for any stream splitting or sample enrichment. Thus, the whole of the eluted sample can be transferred to the source chamber, the only special requirements being a fast scan rate and a means of operating the column outlet at 'zero' pressure. One way of achieving this is to connect the end of the column proper, at 1 atmosphere, to the mass spectrometer through a restrictor consisting of 1 or 2 feet of 0·001-inch tubing. Alternatively, the inlet pressure can be reduced by 1 atmosphere, and the column exit connected directly, i.e. without a restrictor, to the source chamber.

At pumping speeds of less than 100 l/sec, or at source chamber pressures of less than 10^{-4} mm Hg, the maximum allowable inflow rate at atmospheric pressure becomes less than 1 cc/min. For example, the maximum flow rate recommended for the MS9/MS12 type of instrument is 0·15 cc/min. This means that some sort of stream-splitting arrangement or molecular separator is required even with capillary columns. For all mass spectrometers such devices become essential for larger 0·02-inch and 0·03-inch capillaries, including those of the support-coated type, where flow rates upwards of 5 cc/min are used. It should be noted, however, that the greatly increased sample sizes that can be used with the larger diameter columns, especially those coated with solid support, may mean that the 5% or so of the eluate which can be accommodated in the mass spectrometer without an enricher contains much more material than the total amount of eluate which can be handled by a 0·01-inch column. Thus, it may well be that sensitivity is being sacrificed by the use of a thin capillary. Clearly, the column best suited for the GC–MS analysis of a particular unknown mixture should give the maximum possible percentage of eluate in the carrier gas. If supplies of sample in the milligram range are available, then a normal packed column is always preferable from this point of view.

11.4 COMPUTERISED GC–MS

In order to make the best possible use of a fast-scanning GC–MS combination, it is desirable to scan virtually continuously so that several spectra are obtained during the elution of each peak, thereby enabling detection of multi-component peaks. However, the sheer bulk of information which is obtained from a single chromatographic run in the form of MS recorder traces is overwhelming. The situation is suited to computer application at three levels of elaboration of data handling.

In the first place there is the straightforward process of converting the recorded spectra to bar-graph form or equivalent arrays of peak-heights and mass-to-charge ratios, and correlating each spectrum with the corresponding point on the chromatogram. Secondly, there is the comparison of the spectra obtained for individual peaks with previously catalogued spectra of known substances. Thirdly, there is the interpretation of the cracking pattern as indicative of certain functional groups, and even of a complete molecular structure, without direct comparison with standard spectra. Computers have been used with great success for the first two of these procedures but progress has predictably been less spectacular in the third.

11.4a   Data acquisition

Perhaps the simplest way of interfacing a computer with the mass spectrometer is by means of a tape recorder. The recorded analogue signal can subsequently be played back into a digital computer through an analogue-to-digital converter. Such a system has been used for computer processing of high-resolution spectra ($R = 10,000$) obtained at scan rates of 8 seconds per decade.[39] Computer-calculated masses accurate to better than 10 p.p.m. were obtained, which enabled the computer to produce element maps. High-resolution spectra obtained at scan speeds in excess of 40 seconds per decade have been digitised in real time.[56] For low-resolution spectra this is possible even for fast scans of 3 seconds or so. The digitised spectra can be recorded on tape for subsequent batch processing,[57] but a more attractive approach is to feed the data to an on-line computer. Medium-priced computers are available which allow continuous data acquisition

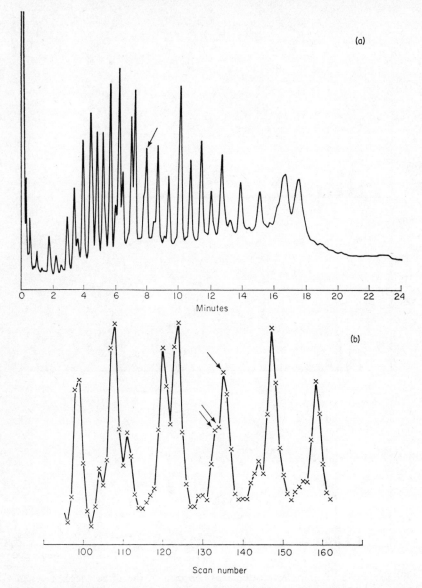

FIGURE 11.14 Illustration of the strong resemblance between (a) a normal flame-ionisation gas chromatogram and (b) a computer-produced plot of total ions collected against scan number. Each cross in (b) corresponds to a stored mass spectrum, and the scans shown cover the period from about 5·5 to 9·5 minutes after injection. Arrows indicate scans for Figure 11.15. (From ref. 58)

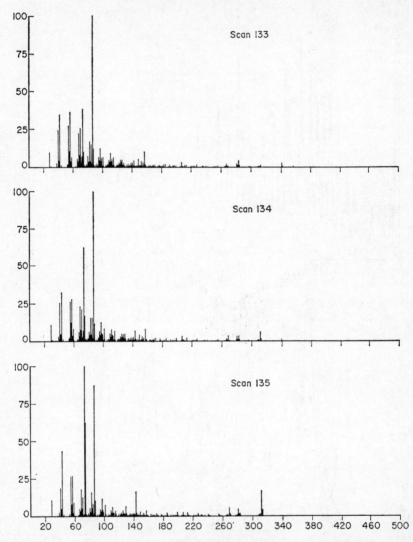

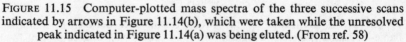

FIGURE 11.15   Computer-plotted mass spectra of the three successive scans indicated by arrows in Figure 11.14(b), which were taken while the unresolved peak indicated in Figure 11.14(a) was being eluted. (From ref. 58)

from analogue inputs to proceed simultaneously with processing of the same data. Thus, the location and intensities of all the MS peaks can be available for storage within milliseconds of the end of a 3-second scan.[58] At the same time, the computer can act as a pseudo-total ion monitor by storing the sum of all the measured ion intensities together with the scan number. A plot of the scanned ion intensity against scan number then resembles the gas chromatogram, and facilitates choice of which spectra should be recalled for inspection, as indicated in Figure 11.14. Figure 11.15 shows three successive spectra obtained from the unresolved peak indicated by arrows in Figure 11.14(b). It can be seen that the middle spectrum is a mixture of the two other spectra, which have been interpreted as methyl 4,8,12,16-tetramethylheptadecanoate and methyl nonadecanoate.[58]

11.4b Library searching

The data acquisition and reduction procedures outlined above take most of the time-consuming labour out of mass-spectral analysis by automatically producing bar graphs or equivalent tabulations. It is at this point that interpretative skill is required. At the present state of the art, such skill remains indispensable for identifying newly discovered or uncommon substances from their spectra. However, it requires only stamina to search the files for a matching spectrum of a known substance. Library searching is an ideal task for a computer, provided that the tolerable variations in relative intensities from identical substances can be well defined. It has been found that computer identification of compounds by comparison of complete spectra is extremely easy.[59a] Such a procedure is, however, wasteful of computer time, and would require an extremely large amount of storage for all catalogued spectra to be stored. Use of the punched-card ASTM compilation of the six strongest peaks for 3200 uncertified mass spectra[21] gives faster, but less certain, identification. Search time can be less than 2 seconds per compound.

11.4c Interpretation

Limited success has been achieved in identifying functional groups from key peaks and characteristic groupings of peaks.[59b] Three methods of assessing spectral type required a total time of 0·9 second. The functional group with the highest probability as determined for the weighted mean of the results of the three methods was found to be correct for 152 compounds out of a library of 182 compounds. This is probably better than the average chemist could achieve.

More positive identification requires the use of additional information. A configuration has been described incorporating UV, IR, NMR, and MS, each with computerised data processing, including the determination of the C, H, and O content of each MS peak. The four sets of analytical data are correlated by a central computer, which determines functional groups and molecular formulae.[60] Were it not for the absence of a gas chromatograph, this system would come close to representing the complete analytical machine.

REFERENCES

1. W. H. McFadden, 'Mass-spectrometric analysis of gas-chromatographic eluents', in *Advances in Chromatography*, Vol. 4 (ed. J. C. Giddings and R. A. Keller), Marcel Dekker, New York (1967), pp. 265–332.

2. J. C. Holmes and F. A. Morrell, 'Oscillographic mass spectrometric monitoring of gas chromatography', *Appl. Spectroscopy*, **11**, 86 (1957). Presented at 4th annual meeting of ASTM Committee E-14, May 1956.

3. R. S. Gohlke, 'Use of a time-of-flight mass spectrometer and vapour-phase chromatography in the identification of unknown mixtures', 132nd meeting of Amer. Chem. Soc., 1957, p. 34B.

4. W. Donner, T. Johns, and W. S. Gallaway, 'Use of a mass spectrometer as a gas chromatograph detector' (radio frequency), reported in *Analyt. Chem.*, **29**, 1378 (1957). Fifth Annual Meeting of ASTM Committee E-14, May 1957.

5. R. S. Gohlke, 'Time-of-flight mass spectrometry and gas–liquid partition chromatography', *Analyt. Chem.*, **31**, 535 (1959).

6. J. H. Beynon, R. A. Saunders, and A. E. Williams, 'Collection of chromatographic fractions in a mass spectrometer sample system', *J. Sci. Instr.*, **36**, 375 (1959).

7. D. Henneberg, 'A continuous procedure for the mass spectrometric analysis of mixtures separated by gas chromatography', *Z. Analyt. Chem.*, **170**, 365 (1959).

8. L. P. Lindeman and J. L. Annis, 'Use of a conventional mass spectrometer as a detector for gas chromatography', *Analyt. Chem.*, **32**, 1742 (1960).

9. A. A. Ebert, 'Improved sampling and recording system in gas chromatography-time-of-flight mass spectrometry', *Analyt. Chem.*, **33**, 1865 (1961).

10. C. Brunnée, L. Jenckel, and K. Kronenberger, 'Continuous mass spectrometric analysis of fractions separated by gas chromatography', *Z. Analyt. Chem.*, **189**, 50 (1962).

11. P. F. Varadi and K. Ettre, 'Operation of the quantitative and qualitative ionization detector and its application for gas chromatograph studies' (radio-frequency device), *Analyt. Chem.*, **34**, 1417 (1962).

12. R. Ryhage and E. von Sydow, 'Mass spectrometry of terpenes', *Acta Chem. Scand.*, **17**, 2025, 2504 (1963); **18**, 1099 (1964).

13. J. A. Dorsey, R. H. Hunt, and M. J. O'Neal, 'Rapid-scanning mass spectrometry. Continuous analysis of fractions from capillary gas chromatograph', *Analyt. Chem.*, **35**, 511 (1963).

14. A. E. Banner, R. M. Elliott, and W. Kelly, 'Use of the mass spectrometer for detection and identification of capillary column effluents', in *Gas Chromatography 1964* (ed. A. Goldup), Institute of Petroleum, London (1965), p. 180.

15. R. Ryhage, 'Use of a mass spectrometer as a detector and analyzer for effluents emerging from high temperature gas–liquid chromatography columns', *Analyt. Chem.*, **36**, 759 (1964).

16. R. Teranishi, R. G. Buttery, W. H. McFadden, T. R. Mon, and J. Wasserman, 'Capillary column efficiencies in gas chromatography—mass spectral analyses', *Analyt. Chem.*, **36**, 1509 (1964).

17. W. H. McFadden and E. A. Day, 'Scan rate considerations in combined gas chromatography mass spectrometry', *Analyt. Chem.*, **36**, 2362 (1964).

18. V. L. Tal'roze, G. D. Tantsyrev, and V. I. Gorshkov, 'Chromato-mass-spectrometry. II. The problems of combining chromatographic columns with a mass spectrometric detector', *J. Analyt. Chem. USSR*, **20**, 91 (1965) [English translation of *Zhur. Analit. Khim.*, **20**, 103 (1965)].

19. R. Ryhage, S. Wilkström, and G. Waller, 'Mass spectrometer used as detector and analyzer for effluent emerging from capillary gas–liquid chromatography column', *Analyt. Chem.*, **37**, 435 (1965).

20. American Petroleum Institute, Research Project 44, and Manufacturing Chemists' Association Research Project, *Mass Spectral Data* tables in loose-leaf form, issued regularly by Chemical Thermodynamics Properties Center, Texas A and M University, College Station, Texas.

21. American Society for Testing and Materials, Committee E-14, *Uncertified Mass Spectra*; six strongest peaks on punched cards: AMD 10.

22. American Society for Testing and Materials, Committee E-14, *Index of Mass Spectral Data*, ASTM Special Technical Publication No. 356, Philadelphia (1963). Also publication AMD 11 (1969).

23. R. S. Gohlke, *Collection of Uncertified Mass Spectra*, Chemical Physics Laboratory, Dow Chemical Co., Midland, Michigan (1963).

24. R. L. Levy, H. Gesser, T. S. Hermann, and F. W. Hougen, 'Application of column bleed absorption in high sensitivity gas chromatography and in gas chromatography–mass spectrometry', *Analyt. Chem.*, **41**, 1480 (1969).

25. R. L. S. Patterson, 'Catty odours in food: confirmation of the identity of 4-mercapto-methylpentan-2-one by GC-MS', *Chem. and Ind.*, **48** (1969).

26. W. J. A. Vandenheuvel and G. W. Kuron, 'Adventitious trimethylsilylation during combined gas chromatography-mass spectrometry', *J. Chromatog.*, **38**, 532 (1969).

27. H. D. Beckey and G. Wagner, 'Analytical uses of a field ion mass spectrometer' (in German), *Z. Analyt. Chem.*, **197**, 57 (1963).

28. J. H. Beynon, *Mass Spectrometry and its Applications to Organic Chemistry*, Elsevier, Amsterdam (1960).

29. J. H. Beynon, R. A. Saunders, and A. E. Williams, *The Mass Spectra of Organic Molecules*, Elsevier, London (1968).

30. K. Biemann, *Mass Spectrometry: Organic Chemical Applications*, McGraw-Hill, New York (1962).

31. H. Budzikiewicz, C. Djerassi, and D. H. Williams, *Interpretation of Mass Spectra of Organic Compounds*, Holden Day, San Francisco (1964).

32. H. Budzikiewicz, C. Djerassi, and D. H. Williams, *Mass Spectrometry of Organic Compounds*, Holden Day, San Francisco (1967).

33. F. W. McLafferty, *Interpretation of Mass Spectra. An Introduction*, Benjamin, New York (1966).

34. R. M. Silverstein and G. C. Bassler, 'Mass spectrometry', in *Spectrometric Identification of Organic Compounds*, 2nd edn., Wiley, New York (1967), Ch. 2.
35. C. C. Sweeley, W. H. Elliott, I. Fries, and R. Ryhage, 'Mass spectrometric determination of unresolved components in gas chromatographic effluents', *Analyt. Chem.*, **38**, 549 (1966).
36. C. Brunnée, L. Jenckel, and K. Kronenberger, 'Use of a mass spectrometer as a specific indicating ionisation detector for gas chromatography' (in German), *Z. Analyt. Chem.*, **197**, 42 (1963).
37. B. H. Kennett, 'Ratio recording of mass spectra in combined gas chromatography-mass spectrometry', *Analyt. Chem.*, **39**, 1506 (1967).
38. A. E. Banner, 'Distortion of peak shape in fast scanning of mass spectra', *J. Sci. Instr.*, **43**, 138 (1966).
39. W. J. McMurray, B. N. Green, and S. R. Lipsky, 'Fast scan high resolution mass spectrometry', *Analyt. Chem.*, **38**, 1194 (1966).
40. J. W. Amy, E. M. Chait, W. E. Baitinger, and F. W. McLafferty, 'A general technique for collecting gas chromatographic fractions for introduction into the mass spectrometer', *Analyt. Chem.*, **37**, 1265 (1965).
41. W. H. McFadden, 'Introduction of gas-chromatographic samples to a mass spectrometer', *Separation Sci.*, **1**, 723 (1966).
42. D. O. Miller, 'High temperature inlet manifold for coupling a gas chromatograph to the time-of-flight mass spectrometer', *Analyt. Chem.*, **35**, 2033 (1963).
43. J. N. Damico, N. P. Wong, and J. A. Sphon, 'Use of activated charcoal to trap gas chromatographic fractions and to introduce volatile compounds into the mass spectrometer', *Analyt. Chem.*, **39**, 1045 (1967).
44. R. P. W. Scott, I. A. Fowlis, D. Welti, and T. Wilkins, 'Interrupted-elution gas chromatography. Its application, with eluate concentration, to the automatic production of simultaneous infrared and mass spectra', in *Gas Chromatography 1966* (ed. A. B. Littlewood), Institute of Petroleum, London (1967), p. 318.
45. C. J. Wolf and J. Q. Walker, 'Pyrolysis gas chromatography combined with interrupted elution for complete analysis of chromatographic effluents', in *Gas Chromatography 1968* (ed. C. L. A. Harbourn), Institute of Petroleum, London (1969), p. 385.
46. J. T. Watson and K. Bieman, 'Direct recording of high resolution mass spectra of gas chromatographic effluents', *Analyt. Chem.*, **37**, 844 (1965).
47. K. Van Cauwenberghe, M. Vandewalle, and M. Verzele, 'Coupling of a gas chromatograph and a mass spectrometer through the direct insertion lock', *J. Gas Chromatog.*, **6**, 72 (1968).
48. W. D. MacLeod and B. Nagy, 'Deactivation of polar chemisorption in a fritted-glass molecular separator interfacing a gas chromatograph with a mass spectrometer', *Analyt. Chem.*, **40**, 841 (1968).
49. M. Blumer, 'An integrated gas chromatograph-mass spectrometer system with carrier gas separator', *Analyt. Chem.*, **40**, 1590 (1968).
50. S. R. Lipsky, C. G. Horvath, and W. J. McMurray, 'Utilization of system employing the selective permeation of helium through a unique membrane of Teflon as an interface for gas chromatograph and mass spectrometer', *Analyt. Chem.*, **38**, 1585 (1966).

51. M. A. Grayson and C. J. Wolf, 'Efficiency of molecular separators for interfacing a gas chromatograph with a mass spectrometer', *Analyt. Chem.*, **39**, 1438 (1967).
52. R. L. Foltz, M. P. Neher, and E. R. Hinnenlamp, 'Reactions of labile trimethylsilyl derivatives with fluorocarbons in a GC-MS system', *Analyt. Chem.*, **39**, 1338 (1967).
53. E. W. Becker, *Separation of Isotopes*, Newnes, London (1961), p. 360.
54. P. M. Llewellyn and D. P. Littlejohn, Pittsburgh Conference on Analytical Chemistry and Applied Spectroscopy, Feb. 1966.
55. D. R. Black, R. A. Flath, and R. Teranishi, 'Membrane molecular separators for gas chromatographic-mass spectrometric interfaces', *J. Chromatog. Sci.*, **7**, 284 (1969).
56. J. R. Chapman, M. Barber, W. A. Wolstenholme, and E. Bailey, 'On-line high resolution mass spectrometry of organic compounds eluted from a gas chromatograph', in *Gas Chromatography 1968* (ed. C. L. A. Harbourn), Institute of Petroleum, London (1969), p. 252.
57. R. A. Hites and K. Biemann, 'A computer-compatible digital data acquisition system for fast-scanning, single-focusing mass spectrometers', *Analyt. Chem.*, **39**, 965 (1967).
58. R. A. Hites and K. Biemann, 'Mass spectrometer-computer system particularly suited for gas chromatography of complex mixtures', *Analyt. Chem.*, **40**, 1217 (1968).
59. L. R. Crawford and J. D. Morrison, 'Computer methods in analytical mass spectrometry', (*a*) 'Identification of an unknown compound in a catalog' and (*b*) 'Empirical identification of molecular class', *Analyt. Chem.*, **40**, 1464 and 1469 (1968).
60. S. Sasaki, H. Abe, T. Ouki, M. Sakamoto, and S. Ochiai, 'Automated structure elucidation of several kinds of aliphatic and alicyclic compounds', *Analyt. Chem.*, **40**, 2220 (1968).

AUTHOR INDEX

This index enables the reader to locate an author's name and work with the aid of the reference numbers in the text. The page numbers are printed in normal type in ascending numerical order, followed by the reference numbers in parentheses. The numbers in *italics* refer to the pages on which the references are listed. This procedure is followed separately for each chapter.

SUBJECT INDEX

Absolute mass detector 137
Absorption, peak abstraction by 73–76
Abstraction techniques 66–83
  complete removal 66–79
  pre-column partition 79–83
  with regeneration 68–70, 75
Abstractors 66–83
  chemical 76–79
  physical 73–76
Acetates 51, 52, 62, 98
Acetone, see Ketones
Acetylation 101
Acetylene, see Alkynes
Acids (carboxylic) 53, 67, 75, 78
  abstraction of 67, 78, 93
  reactions of 99, 104
  see also Amino-acids, Bile acids
Activity coefficient 10–20
Adjusted retention volume, V_R' 7
Alcohols 26, 31, 32, 47, 51, 52, 61, 62, 67, 74, 77, 78, 88, 90, 91, 98, 99, 104, 128, 130
Aldehydes 61, 62, 67, 75, 78, 87, 88, 90, 91, 95, 104, 130, 252
Aldol condensation, on molecular sieve 75
Alkadienes, see Dienes
Alkaloids 96
Alkanals, see Aldehydes
Alkanes 18, 19, 26, 28, 32, 33, 43, 46, 48, 49, 50, 52, 55, 58, 62, 67, 71, 72, 75, 76, 91, 102, 104, 128, 167
Alkanols, see Alcohols
Alkenes 19, 24, 26, 29, 50, 55, 62, 66, 67, 71, 72, 74, 76, 90, 91, 102, 121, 128
Alkyl halides 67, 78, 79
Alkylsilanes 43
Alkynes 50, 55, 67, 71, 74, 76
Alumina modified with salts, use as abstractor 67, 74
Aluminium 169
Amides 104

Amines 88, 100, 104, 137
  aromatic 67, 78
Amino-acids 95, 100
Antoine equation 10
Apiezon 26, 35, 50, 56, 62
Argon β-ionisation detector 182
Aromatic hydrocarbons 26, 30, 32, 34, 36, 50, 53, 67, 71, 74, 88, 91, 118
  see also Benzene, Phenols, Polycyclic hydrocarbons
Arsenic compounds 155

Bacteria, analysis by PGC 118
Band pass filters for fast-scanning IR 220
Beam condenser, use in GC–IR 212
Beer–Lambert law 211
Beilstein detector 169–170
1,2-Benzanthracene 167
Benzene 18, 24, 32, 33, 34, 73, 75, 153
Benzidine 67, 78
Bile acids 99
Biological material, analysis by PGC 118
Boiling point, correlation with retention 47–50
  relation to vapour pressure 10, 18
Boric acid, abstraction by 67, 77
Bromine, see Halogen compounds
Bromine water, unsuitability for on-column abstraction 76
Butyrates 62
  see also Steroids
Butyrolactone 22

Caesium bromide 155
Capillary column 216, 217, 255
Carbohydrates 95, 98
Carbon dioxide as carrier gas 139, 206
Carbon number, correlation with retention 51–54
  see also Retention index
Carbon skeleton chromatography 61, 103–105

275